M.

This b
and n.

Surface Engineering Casebook

Solutions to corrosion and wear-related failures

Edited by J S Burnell-Gray and P K Datta

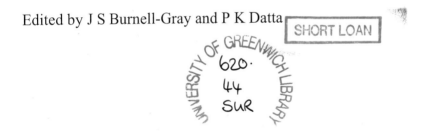

WOODHEAD PUBLISHING LIMITED

Published by Woodhead Publishing Limited,
Abington Hall, Abington, Cambridge CB1 6AH, England

First published 1996, Woodhead Publishing Limited

British Library Cataloguing in Publication Data
A catalogue record for this book is available from the British Library.

ISBN 1 85573 260 2

Printed by St Edmundsbury Press, Bury St Edmunds, Suffolk, England

Contents

List of contributors

Dr J S Burnell-Gray
School of Engineering
University of Northumbria at Newcastle
Newcastle upon Tyne
NE1 8ST

Prof P K Datta
School of Engineering
University of Northumbria at Newcastle
Newcastle upon Tyne
NE1 8ST

Chapter 1

Prof A Matthews
Research Centre in Surface Engineering
University of Hull
Cottingham Road
Hull
HU6 7RX

Chapter 2

Dr P A Gillespie
Centre for Quantum Metrology
National Physical Laboratory
Teddington
Middlesex
TW11 0LW

Chapter 3

Mr K Harrison
Sulzer Metco UK Ltd
Westmead
Farnborough
Hants
GU14 7LP

Chapter 4

Prof D Kirkwood
Head of Department of Engineering
Glasgow Caledonian University
Glasgow

Chapter 5

Prof Dr W Funke
c/o Forschungsintitut für Pigmente und Lacke et V
Allmandring
D-7000 Stuttgart 80 (Vailhingen)
Germany

Chapter 6

Dr K Natesan **Chapter 7**
Materials & Components Technology Division
Argonne National Laboratory
9700 South Cass Avenue
Argonne
Illinois 60439
USA

Mr R Wing **Chapter 8**
CUK Ltd
Bramble Way
Clover Nook Industrial Estate
Somercotes
Alfreton
Derbyshire
DE55 4RH

Dr D E Taylor **Chapter 9**
School of Engineering & Advanced Technology
University of Sunderland
Chester Road
Sunderland
SR1 3SD

Prof D T Gawne **Chapter 10**
School of Engineering Systems & Design
South Bank University
Borough Road
London
SE1 0AA

Dr C Subramanian **Chapter 11**
Surface Engineering Research Group
Department of Metallurgy
Gartrell School of Mining, Metallurgy & Applied Geology
The University of South Australia
Adelaide, SA5095
Australia

Dr N Whitehouse **Chapter 12**
Paint Research Association
Waldegrave Road
Teddington
Middlesex
TW11 8LD

Dr A Swift **Chapter 13**
CSMA Ltd
Armstrong House
Oxford Road
Manchester
M1 7ED

Chapter 1

Introduction

1.1 Scope of the Casebook

This Casebook has been prepared focusing attention on current and near-future surface engineering practice. The structure and contents of the book have been conceived to provide a number of interrelated themes and a coherent philosophy. The text provides a useful blend of theory, problem identification, range of possible solutions and rationale for the selection of a particular surface engineered solution to combat corrosion and wear.

Surface Engineering Casebook comprises 13 chapters as indicated below. *Chapters 2, 3* and *4* consider surface engineering technologies – here there is a general emphasis on the PVD processes. *Chapters 5* to *11* concern corrosion and wear. *Chapter 12* gives an appreciation of paint technology. Finally, *Chapter 13* deals with surface analysis and characterisation – areas central to surface engineering and holding particular promise for improvements in existing and emerging surface engineering techniques.

In the Introduction an attempt is made to review the state of the art in this field, describe some recent developments and industrial applications, and identify future trends in surface engineering applied to the control of corrosion and wear.

1.2 Case studies

Case studies describe aspects of real problems in sufficient detail for the engineer to understand the corrosion/wear situation, analyse and evaluate the range of potential solutions, and make decisions relating to a recommended course of action. Furthermore:

- Each case study is based upon a real corrosion or wear failure and is designed to illustrate corrosion and wear mechanisms as well as surface engineering principles and to provoke discussion rather than to reflect good or bad engineering practice.

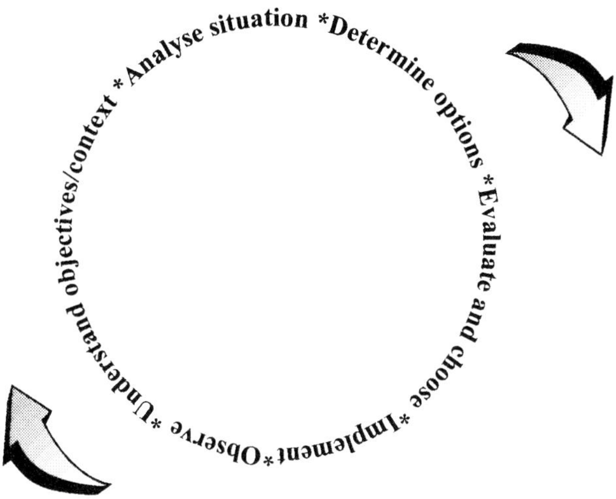

Figure 1.1 The stages of case study analysis [1].

- The case study is a way of simulating some aspects of practical engineering decision-making.
- Case studies stimulate the acquisition and development of analytical skills. The process of logic and clarification the reader goes through to understand what is critical in a case study matches closely real engineering decision-making.
- Cases provide an appropriate method for validating theories and examining how they fit in real engineering situations [1].

Corrosion and wear management cases often reflect the complexity of the failures themselves. Interacting within any decision-making scenario are diverse degradation mechanisms, types of technology, and systems and procedures [1].

Case studies can suggest that the reader does several things – usually they require an understanding of the failure and its implications as well as the solution to specific problems. A particularly useful approach to analysis is to consider cases as problem-solving exercises and follow the sequence of activities illustrated in Fig. 1.1, viz.:

1. The mechanisms leading to corrosion and wear failure must be recognised.
2. The overall objectives of the operation and problem-solving process must be understood.
3. The nature of the problem should be analysed and the interrelationships between different parts of the case established. By this time the overall structure

of the problem should be clear and then the engineer progresses to consider the different options to improve corrosion and wear control.

4. Eventually it will become necessary to evaluate and make recommendations.
5. After this the recommended solution will be implemented.
6. The effectiveness of the implemented solution should be observed and if any further action is needed the whole cycle is repeated [1].

1.3 Surface engineering – background

Surface engineering is an enabling technology applicable to a wide range of industrial sector activities [2–6]. It encompasses techniques and processes capable of creating and/or modifying surfaces to provide enhanced performance such as wear, corrosion and fatigue resistance, and biocompatibility – Fig.1.2 and Tables 1.1 to 1.3. The newer surface engineering techniques together with the traditional surface treatments have a profound influence on several engineering properties, as can be seen in Fig. 1.2. Surface engineering processes can now produce multilayer and multicomponent surfaces, graded surfaces with novel properties and surfaces with highly non-equilibrium structures. Datta et al. [2–4] have identified, in a broad sense, three interrelated activities describing surface engineering:

I. Optimisation of surface/substrate properties and performance in terms of corrosion, adhesion, wear and other physical and mechanical properties.
II. Coatings technology including the traditional techniques of painting, electroplating, weld surfacing, plasma and hypervelocity spraying, various thermal and thermochemical treatments such as nitriding and carburising, as well as the newer combinations of laser surfacing, physical and chemical vapour deposition, ion implantation and ion mixing.
III. Characterisation and evaluation of surfaces and interfaces in terms of composition and morphology, and mechanical, electrical and optical properties [2–4].

Surface engineering is a rapidly evolving discipline which enables the design and manufacture of metallic, ceramic, polymeric and composite systems, with unique combinations of bulk and surface properties obtainable in neither the substrate nor the surface material alone. The principal reason for the growing importance of this technology is that the degenerative processes – as illustrated in Table 1.1 – in most technological applications focus at the surface of a component, thereby requiring surface properties differently designed to the base [5]. This applies to a wide range of engineering applications, as exemplified by the wear coatings listed in Table 1.4.

It is now widely recognised – see for instance the applications cited in Table 1.4 – that the successful exploitation of these processes and coatings may enable

Table 1.1 Typical causes of failure of components and devices, and applications of surface engineering [5]

Failure	Applications
Electronic	
• Contamination by certain impurities	* Ion implantation for semiconductor devices
• Diffusion/segregation of constituent materials in semiconductors	* Metal-silicide layers for interconnector strips
	* Thermo-junctions
	* Superconducting films
	* Well-bonded contacts with linear (ohmic) properties
Mechanical	
• Wear, seizure, fusion of surfaces in contact	* Hard layers for cutting tools/drills
• Hydrogen diffusion, embrittlement	* Low friction layers for bearings
• Phase change, precipitation, diffusion/segregation of constituent materials, amorphous/crystalline transition	
Chemical	
• Dissolution of layers	* Corrosion resistant layers
• Pitting	* Catalytic layers
• Segregation/diffusion of constituent materials	* Electrodes for gas sensors
• Oxidation and other reactions	
• Contamination by active species	
Physical	
• Adhesion failure, flaking, blistering	* Magnetic layers for recording tapes and discs
• Phase changes	* Diffusion barriers, e.g. to protect food
• Contamination, such as formation of impurities in optical devices	* Coatings for optical lenses
• Density/refractive index changes	
• Diffusion/segregation	

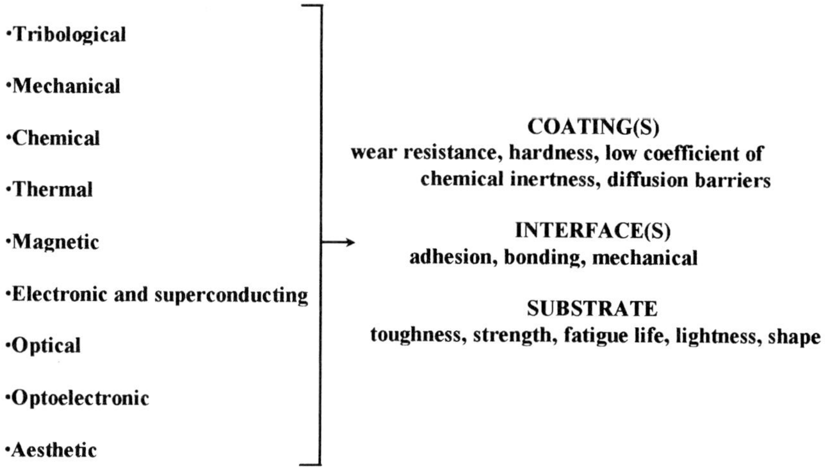

Figure 1.2 Design attributes and selection criteria [5].

Table 1.2 Properties required by engineering components [7]

1. Abrasive wear resistance under conditions of low/high compressive loading
2. Resistance to scuffing or seizure
3. Bending or torsional strength
4. Bending or torsional fatigue strength
5. Resistance to mechanical pitting (surface contact fatigue)
6. Resistance to case crushing (surface collapse)
7. Resistance to corrosion and erosion

the use of simpler, cheaper and more easily available substrate materials, with substantial reduction in costs, minimisation of demands for strategic materials and improvement in fabricability and performance: many of these techniques, especially PVD, can be carried out in an environmentally friendly manner. In demanding situations where the technology becomes constrained by surface-related requirements, the use of specially developed coating systems may represent the only real possibility for exploitation [2–4].

Surface engineering technologies span five orders of magnitude in thickness and three orders of magnitude in hardness. The thickness of the engineered surface can vary from several millimetres for weld overlays to a few micrometres – or even nanometres – for physical vapour deposited (PVD) and chemical vapour deposited (CVD) coatings, while the depth of surface modification induced by ion implantation is $\leq 0.1\,\mu m$. Similarly, examples of coating hardnesses are cited by Sankaran [5]: 250–300 H_V for some spray coatings; 1,000 H_V for nitrided steels; 1,300–1,600 H_V for detonation gun (D-gun) carbide in metal cermet

Table 1.3 Wear processes and wear resistant thermally sprayed coatings [8]

Wear process	Typical thermally sprayed coatings used
Abrasion	Al_2O_3, ZrO_2, Cr_2O_3, NiCrBSiC, WC–Co, TiC–Ni, Cr_3C_2–NiCr
Cavitation erosion	NiTi, Cu–Ni, NiCrBSiCAlMo, 316S/S
Liquid impingement erosion	Stellite, Al_2O_3, WC–Ni
Fretting	
(low amplitude)	Cu–Ni, Cu–Ni–In
(high amplitude)	CoMoCrSi, Mo–NiCrBSi, T–800
Abradable	CoNiCrAlY, Ni–graphite, NiCrAl–bentonite
Impact and sliding	WC–Co, T–800, Cr_3C_2–NiCr
Galling	Cr_3C_2–NiCr, T–800
Solid particle erosion	WC–Co, Cr_3C_2–NiCr, WC–NiCrBSiC, Cr_3C_2–MCrAlY, T–800
Adhesive	Al–bronze, Mo–NiCrBSi, Al_2O_3+TiO_2, WC–Co
Scuffing	Mo, Mo–NiCrBSi
Dry sliding	WC-Co, Cr_2O_3, Cr_3C_2–NiCr, NiCrBSi
Biological implants	TiO_2, ZrO_2, hydroxyapatite, fluorapatite

Table 1.4 Industries and components using thermally applied wear coatings [8]

Flight gas turbines	Land-based turbines	Others
Turbine and compressor blades, vanes	Turbine and compressor buckets, vanes, nozzles	Feed rolls Pump sleeves
Gas path seals	Piston rings (IC engines)	Shaft sleeves
Mid-span stiffeners	Hydroelectric valves	Gate valves, seats
Z-notch tip shroud	Boiler tubes	Rolling element bearings
Combustor and nozzle assemblies	Wear rings Gas path seals	Dies and moulds Diesel engine cylinder
Blade dovetails	Impeller shafts	Hip joint prostheses
Flap and slat tracks	Impeller pump housings	Hydraulic press sleeves
Compressor stators		Grinding hammers Agricultural knives Biological implants

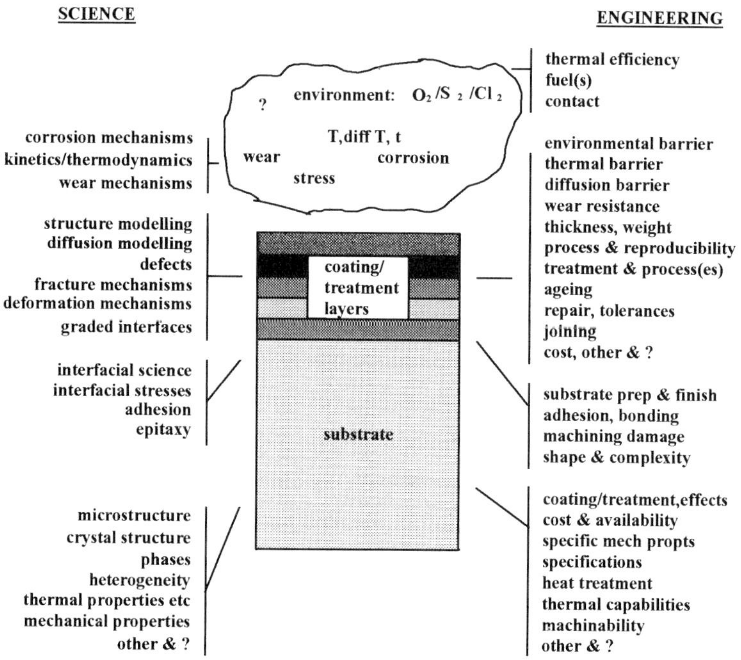

Figure 1.3 Aspects of surface engineering [4].

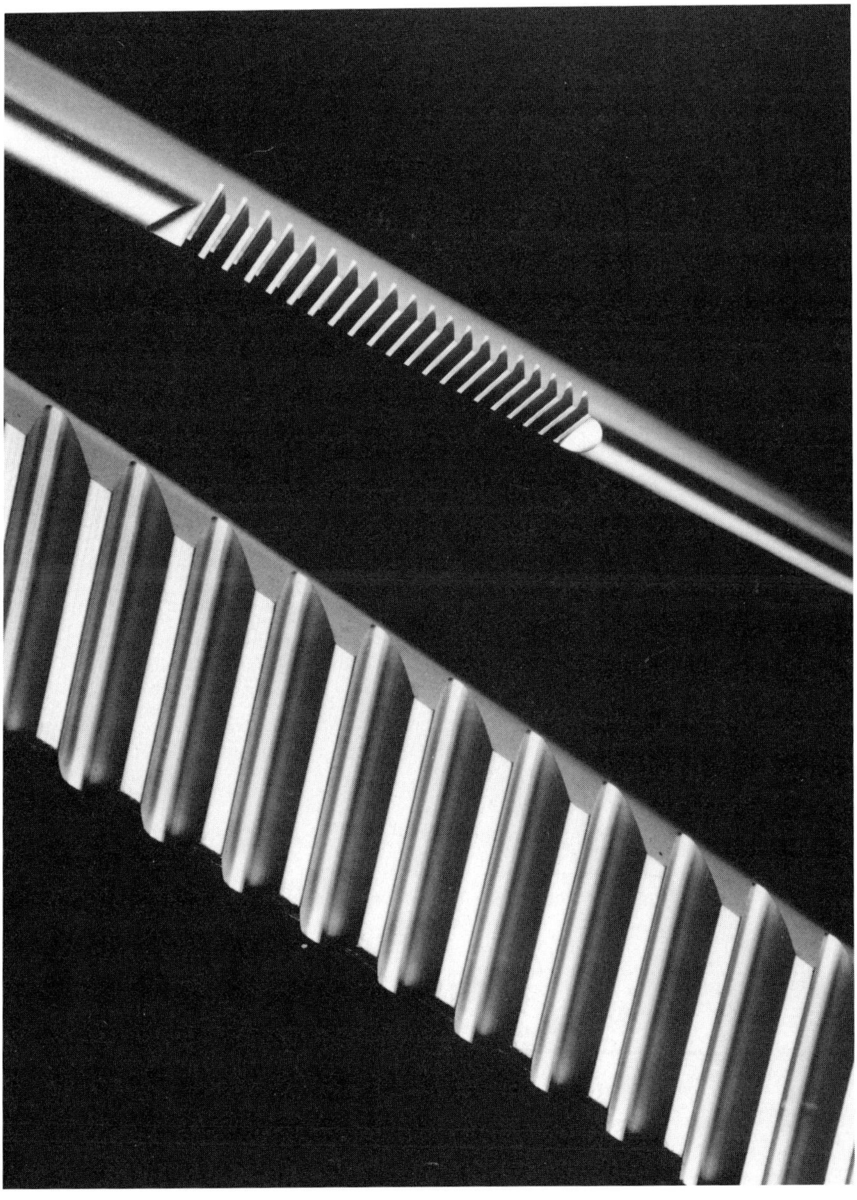

Figure 1.4 TiN coated brooches – courtesy of Multi-Arc (UK) Ltd.

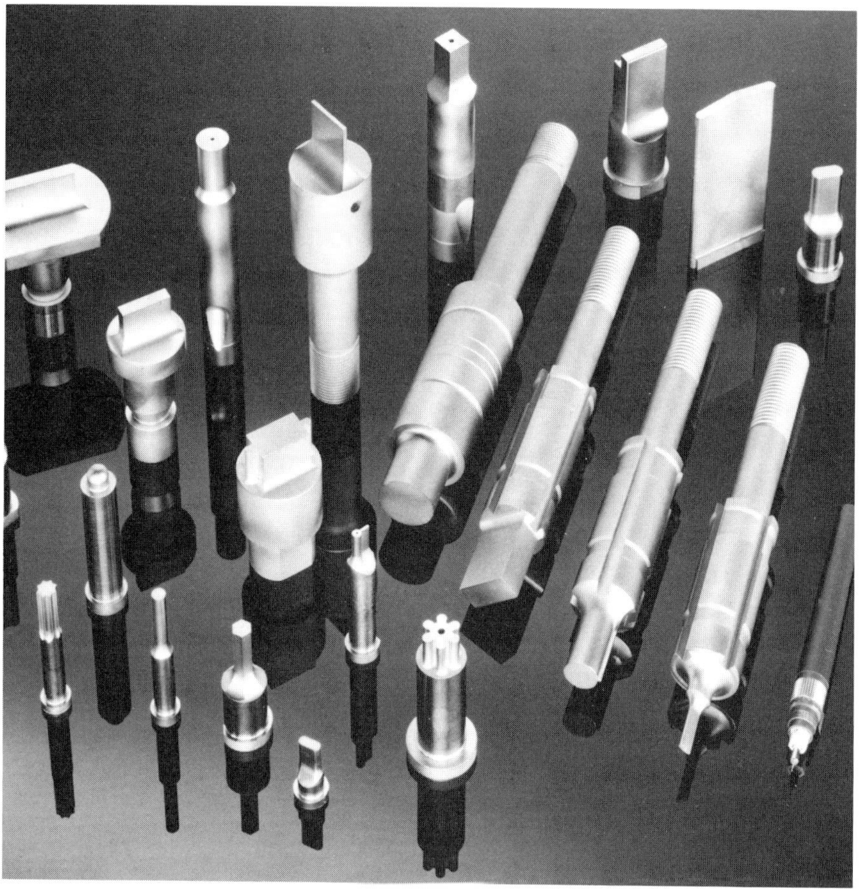

Figure 1.5 TiN coated components – courtesy of Multi-Arc (UK) Ltd.

coatings; 3,500 H_V for titanium nitride PVD coatings; and ≤10,000 H_V for diamond coatings.

Each surface engineering technique has advantages and limitations, which must be evaluated for a specific application. The choice of material and treatment is dictated by three zones of interaction in the system: the substrate, interfaces and surface, as indicated schematically and for actual components in Fig. 1.3 to 1.6 [5,6].

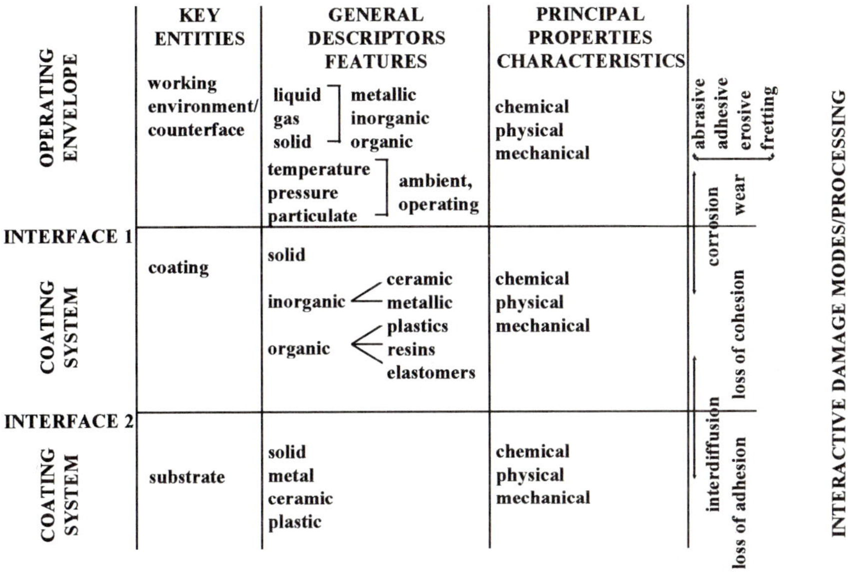

OPERATING ENVELOPE
INTERFACE 1
COATING SYSTEM
INTERFACE 2
COATING SYSTEM
KEY ENTITIES
GENERAL DESCRIPTORS FEATURES
PRINCIPAL PROPERTIES CHARACTERISTICS
working environment/ counterface
liquid
gas
solid
metallic
inorganic
organic
temperature
pressure
particulate
ambient, operating
chemical
physical
mechanical
coating
solid
inorganic
ceramic
metallic
organic
plastics
resins
elastomers
chemical
physical
mechanical
substrate
solid
metal
ceramic
plastic
chemical
physical
mechanical
abrasive
adhesive
erosive
fretting
wear
corrosion
loss of cohesion
interdiffusion
loss of adhesion
INTERACTIVE DAMAGE MODES/PROCESSING

Figure 1.6 Generalised features of a working coating system [6].

1.4 Corrosion and wear-related failures

Surface engineering techniques generally consist of surface treatments where the composition/structure or the mechanical property of the existing surface is modified, or a different material is deposited to create a new surface. Tables 1.5–1.7 indicate how surface engineering is essential in the application and exploitation of high performance engineering components. This is especially true in relation to both the rising costs of advanced performance structural materials and the increasingly high life-cycle costs associated with high performance systems.

Comparative costs of various processes used to impart wear resistance are listed in Table 1.8. Whilst Table 1.9, derived from Sankaran [5], further develops the economics aspect by citing the relative activity of markets for thermal spray coatings.

Table 1.5 Key users of surface finishing [9]

	Principal coatings (£M)			Total	Share
Industry	**Organic**	**Plating**	**Galvanising**	**(£M)**	**(%)**
Motor vehicle	308	170	13	491	28
Construction	324	5	123	452	26
Electrical/electronic	109	230	-	339	20
Retail consumer	138	150	18	306	18
Aerospace	59	73	-	132	8

Table 1.6 Surface finishing – value by industry sectors [9]

Coating	Size (£M)	Share (%)
Organic	1,450	43
Plating	705	21
Galvanising	355	10
Surface heat treatment	325	10
Hard facing	100	3
Anodising	65	2
Tin plating	40	1
Vitreous enamelling	40	1
PVD & CVD	25	1
Others	250	8
Total	3,355	100

1.5 Surface engineering technologies

The basic features of a wear or corrosion resistant coating system can be described with reference to Fig. 1.6. The coating system – comprising coating, interfaces and substrate – is required to respond appropriately to an operating environment or counterface. Interaction across *Interface 2* may lead to performance failure, demonstrated by unacceptable and excessive interdiffusion: during fabrication (e.g. in diffusion coating by chromising or aluminising), during elevated temperature service experienced by gas turbine blades, or following fast cutting

Table 1.7 Thermal spray coating activity by industry [5]

Sector	Current activity	Growth potential to 2000
Aircraft gas turbine	1	1
Industrial gas turbine	4	2
Steam turbine	4	3
Diesel engines	3	2
Automotive engines	5	1
Transportation	4	2
Chemical processing	4	2
Oil and gas exploration	3	3
Paper and pulp	3	3
Electric utility	2	3
Textiles	3	5
Electric/electronic	4	2
Medical/dental	4	5
Iron and steel making	3	5
Business equipment	4	2
Defence and aerospace	4	3

Notes: Scale based on 1 equal to highest growth and 5 equal to modest growth
For comparison purposes, current activity of 5 or growth potential of 5 is higher than any of the remaining industries not listed here

Table 1.8 Economic comparisons of selected surface technologies for wear applications [10]

Process	Fixed costs	Variable costs	Profitability
Nitriding			
Salt bath	30	200	100
Gas	80	120	100
Plasma	100	100	100
Ion implantation	500	100	100
Coating			
Galvano-techniques	30	200	100
CVD	100	150	300
PVD	400	300	500
PACVD	200	150	500

or forming operations.

The damage modes occurring across *Interface 1* are generally recognised as corrosion or wear. Corrosion takes place either at ambient temperatures (often in aqueous environments – *wet* corrosion) or at elevated temperatures (often termed *dry* corrosion). Wear processes are diverse and complex but can be classified – depending on the nature of the environment/counterface – as *abrasive*, *adhesive*, *fatigue* or *corrosive*.

Table 1.9 Estimated current and future powder consumption by type [5]

Powder type	Powder consumption (10^3 lbs/year) Current	Year 2000
Cobalt-cemented tungsten carbide [1]	600–650	1,000–1,100
Chrome carbide	65–75	175–200
Stabilised zirconia	525–600	2,300–2,500
High purity alumina	175–200	675–700
Alumina/titania	150–175	375–400
Chromia [2]	200–225	350–375
Other superalloys	125–150	No projections
MCrAlY alloys	325–375	600–625
Molybdenum blends [3]	275–300	575–615
Special composites (compressor abradable powders)	400–460	800–850
Fluxing alloys [4]	475–500	Moderate/no growth
Non-ferrous alloys	240–250	Moderate growth
Iron, nickel and cobalt alloys	775–825	2,000–2,200

Notes: [1] includes approximately 50,000 lb self-fluxing powder as blend
[2] does not include internal consumption by Union Carbide Coating Services
[3] includes average 20% by weight fluxing alloy powder
[4] does not include projections for powder used in blends noted in [1] and [3]

Table 1.10 Surface modification and coating techniques [after 5 and 13]

Surface treatments	Overlay coatings
Thermal treatments	Plating
* induction hardening	♦ electroplating
* flame hardening	♦ mechanical plating
* laser hardening	
* electron beam	
Thermochemical diffusion treatments	Weld cladding
* carburising	♦ oxy-acetylene
* nitriding	♦ tungsten inert gas
* carbonitriding	♦ metal inert gas
* chemical – etching, oxidation	
Mechanical treatments	Thermal spraying
* shot-peening	♦ flame
* grinding	♦ detonation (D-) gun
* cold working	♦ plasma
Ion implantation	Chemical vapour deposition (CVD)
Laser glazing	Physical vapour deposition (PVD)
	♦ sputtering
	♦ evaporation

Coating	Example
Thin films and coatings	• Chemical (conversion coatings) • Electro deposition • Electroless deposition • Thermal spraying – flame, plasma, arc • PVD – evaporation, sputtering, laser ablation, ion beam-assisted deposition • CVD – plasma-assisted CVD (PACVD)

Total incompatibility between the coating and the environment/contacting face is required. In contrast good compatibility between a coating and substrate is necessary to create an adherent coating system. Since these fundamentally different requirements have to be met across *Interfaces 1* and *2*, no single coating can simultaneously satisfy such criteria, and a compromise is necessary in practical systems. Even so, adequate systems can be developed. However, modern advanced process technologies, or hybrids, offer the prospect of greatly improved and optimized performance [11].

Strafford and co-workers [11] have listed the characteristics and properties necessary to fully define a coating, viz.: density, composition and stoichiometry, structure, morphology and microstructure, surface finish, thickness, adhesion, cohesion, hardness, internal stress, strength, fracture toughness, ductility, elastic modulus, Poisson's ratio, friction coefficient, thermal expansion coefficient, specific heat, stress-temperature behaviour and thermal conductivity.

It was only in the 20th century that new surface engineering techniques – Table 1.10 – could be exploited due to rapid improvements in various technologies.

Table 1.11 Coatings produced by PVD [14]

Coating	Application (colour)	Deposition temperature (°C)
TiN	Tools, machine and decorative parts (gold)	400–500
TiC_3N_7	Forming tools, machine parts (gold)	350–400
TiAlN	For extreme operational conditions: cutting tools, machine and decorative parts (dark violet)	700
Cr	Strongly bonded coating for wear resistant depositions on non-ferrous metals and temperature sensitive machine parts (metallic)	600
CrN	Machine parts for low-temperature materials (metallic to grey)	200–350

More recently growing commercial maturity of the semi-conductor industry has spawned many gains from advances in surface science and surface engineering such as PVD – Tables 1.11 and 1.12 – laser and electron beam processing, ion implantation – Table 1.13 – and plasma thermochemical techniques.

Deposition procedures, include traditional electrodeposition and chemical conversion coating, together with newer methods such as thermal spraying – where a plasma or electric arc is used to melt a powder or wire source, and droplets of molten material are sprayed on to the surface to produce a coating – see Tables 1.14 and 1.15; PVD, in which a vapour flux is generated by evaporation, sputtering or laser ablation; and CVD, where reaction of the vapour phase species with the substrate surface produces a coating.

Surface treatments include:

Table 1.12 Attributes of PVD titanium nitride coatings [5]

- Excellent adhesion due to the high arrival energy of the coating material and the ability to thoroughly sputter clean the surface prior to coating
- Uniform thickness due to gas scattering and the ability to rotate component
- Surface finish that in the better systems equals that of the substrate, eliminating finish machining
- No process effluents or pollutants
- No hydrogen embrittlement
- Dense structures
- Controllable and repeatable stoichiometry and crystallographic structure
- Wide range of coatings and substrate materials – metals, alloys and ceramics
- Multiple coating possible
- Low temperatures (< 500°C)
- Greater productivity and major cost savings

Table 1.13 Industrial exploitation of ion implantation [15]

Material	Application (specific examples)	Typical results
Cemented WC	Drilling (printed circuit board, dental burrs etc.)	Four times normal life, less frequent breakage and better end product
Ti-6Al-4V	Orthopaedic implants (artificial hip and knee joints)	Significant (400 times) lifetime increase in laboratory tests
M50, 52100 steel	Bearings (precision bearings for aircraft)	Improved protection against corrosion, sliding wear and rolling contact fatigue
Various alloys	Extrusion (spinnerets, nozzles and dies)	Four to six times normal performance
D2 steel	Punching and stamping (pellet punches for nuclear fuel, scoring dies for cans)	Three to five times normal life

* mechanical processes that work-harden the surface – e.g. shot-peening;
* thermal treatments which harden the surface by quenching constituents in solid solution – e.g. laser or electron beam heating;
* diffusion treatments which modify the surface composition – e.g. carburising and nitriding;
* chemical treatments that remove material or change the composition by chemical reactions – e.g. etching and oxidation; and
* ion implantation where the surface composition is modified by accelerating ions to high energies and implanting them in the near-surface [5].

Comparisons of certain of the above techniques are contained within Table 1.16 and examples of deposition and treatment technologies used in the aerospace industry are given in Table 1.17.

1.6 Surface engineering – state of the art and future developments

1.6.1 State of the art

Since the early 1980s there has been a continuing and rapid development of

Table 1.14 Thermal spray materials for protection and repair [16]

Tribological property	Material
Protection from:	
* abrasion	Tungsten carbide, cobalt
* fretting	Copper, nickel, indium
* abradability	Aluminium, silicon, polyesters
* erosion	Chromium carbide, nickel, chromium
Repair	Nickel, aluminium

Table 1.15 Examples of thermal spraying applications [17]

Turbocharger shafts	Rotating at up to 100,000 rpm against floating bronze bearings. Worn journals are reclaimed by applying a plasma sprayed coating of a heat and wear resistant material.
Orthopaedics	Increasing reliance on the use of plasma spraying for coating artificial joints. The long-term stability of these prostheses has been enhanced by the use of inert coatings with controlled porosity into which bone may grow to provide a strong fixation. Further developments include the spraying of bio-active ceramics which also promote bone growth.
Pierre Laporte Bridge	This suspension bridge, which spans the St Lawrence River near Quebec Canada, was opened in 1970 and at the time was protected by a paint system. However, the maintenance cost was such that after six years the structural girder work was grit-blasted to bare metal and sprayed, in situ, with a 150μm thick coating of zinc. The coating was sealed with two layers of vinyl sealer.
Cylinder bore	In service a worn bore can be reclaimed by pre-machining and subsequently plasma sprayed with a special wear and corrosion resistant coating. It can then be finished by machining to restore it to original size and surface finish.
Gear synchronising cone	In 1960 a major automobile manufacturer decided to improve the synchro mesh on gear boxes fitted to automotive engines. After testing, a sprayed coating of molybdenum on the gear cone was specified to prevent pick-up caused by the friction between the cone and the female synchroniser. This practice is still in operation after 35 years.

advanced surface engineering practices for the optimisation of corrosion and wear resistance. It is now possible to produce coatings of novel composition and microstructure in multilayer/multicomponent format – Fig. 1.7 – as appropriate to the design audit, by a variety of sophisticated physical and chemical processes, including hybrid technologies.

As discussed earlier there is a paradox concerning compatibility between the environment, coating and substrate. In principle, no single coating can satisfy the totally different compatibility demands at the environment/coating interface, and those of the coating/substrate. Surface conditions, cleanliness and finish are also of major significance – see Fig.1.8 for instance.

This paragraph uses *shot-peening* as an example of the state of the art in surface engineering. Shot-peening is a well established surface treatment which, when

Table 1.16 Comparison between five surface engineering processes [17]

Process	Resistance to wear	Risk of distortion	Resistance to impact	Convenience	Range of materials	Typical thickness
Plasma spraying – atmospheric	High	Low	Low	Very good, gun is offered to the work	Extensive	0.025 to 3 mm
Plating	High	Low	Medium	Low, work is processed in a bath	Low	≤ 0.01 mm
Welding	Medium	High	Good	Good	Medium	> 3 mm
CVD/PVD ion deposition	High	Low	Good	Low vacuum chamber required	Good	Thin film, micron size
Cladding	Low	Low	Good	Good	Low	> 3 mm

Table 1.17 Surface engineering technologies used in the aerospace industry [18]

Technique	Material	Requirement	Application
Mechanical treatments, e.g. peening	Steels, titanium-based and nickel-based alloys	Improved mechanical and wear properties	Compressor blade roots
Paints	Phenolic and epoxy polyurethanes	Cosmetic, corrosion and wear, earthing, emissivity and infrared	Shafts, discs, blading
Polishing	Steels, titanium-based and nickel-based alloys	Cosmetic, salvage and repair efficiency	Aerofoil surfaces on vanes and blades
Electrochemical	Tribomet, chromium	Corrosion and wear, salvage and repair	Bearing chambers, stator vanes
Thermal spraying (D-gun, flame spraying, plasma spraying)	Al/Si polyester, WC/Co, CuNiIn	Corrosion and wear, salvage and repair, seals, net-shapes	Snubbers, gas-path seals, combustor cans
Thermochemical	Nitrogen and carbon into steels	Improved mechanical properties	Shafts and gears
Pack aluminising	Nickel-based alloys	Corrosion/oxidation	Aerofoil surfaces on vanes and blades

carried out in a controlled manner and specification, can markedly increase a component's life by modifying the undesirable stress patterns induced during machining or forming. Finishing operations such as grinding often result in residual tensile or compressive surface or sub-surface stresses, which are responsible for considerable variations in fatigue strength. Here, shot-peening

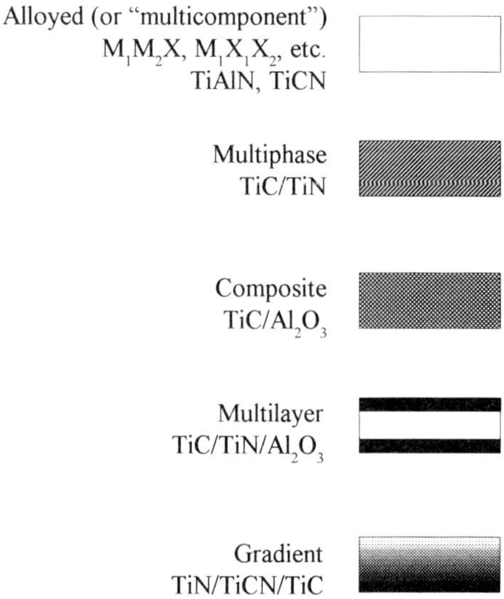

Figure 1.7 Schematic of the structure of different types of coatings [6].

of components after finishing introduces compressive stresses which greatly improve fatigue properties. For instance in a steel of hardness 50 HRC, peening increases the fatigue strength to almost twice that of smooth unpeened material, and almost four times cf. notched unpeened material. In components such as gears or aerospace parts made from high hardness materials, notch sensitivity may be significant. In such cases, shot-peening can be used to introduce a high degree of damage tolerance. Recent developments in *duplex* or *triplex* peening are now used on sophisticated machined parts. To achieve maximum surface compressive stress duplex shot-peening is used, in which the first peening is performed with large shot, to the depth of the residual compressive stress followed by similar coverage with small shot to induce a high compressive residual surface stress and an increase in fatigue properties. So much so that duplex shot-peening is often the only way to impart fatigue strength in a critical component without expensive redesign. Another area where shot-peening is being increasingly applied is in the production of aircraft wings, fuselage and door panels [5] – Table 1.17.

1.6.2 Future developments

At the research level, scientists and engineers need to acquire a much better systematic understanding of the various process technologies. In the PVD processes, one such issue – which offers a potent competitive advantage over more conventional treatments/processes – concerns plasma densities and their significance in allowing the repeatable creation of advanced surface engineered artefacts with outstanding properties and performances. A further critical issue is the precise significance of measured properties and characteristics – e.g. hardness – in relation to actual coating performance. Here insight is needed into the consequence of particular hardness levels in relation to wear performance, so that a coating engineered to a given hardness could be expected to offer a particular design wear life.

Also at the fundamental level there is a need to understand structure/property relationships in, for example coatings, so that surface engineered systems can be designed from conception to develop desired properties. In this regard there is considerable scope for the creation of tailored coatings of chosen composition, structure and properties – including multilayer/multicomponent format – by highly adaptable PVD and CVD technologies.

Strafford et al. [11] believe that considerable progress could be made in this area of design by analysing the performance of proven coating systems, for example in multilayer coated tools. Thus the rôle of individual layers in a multilayer coating in resisting wear and failure at particular parts of a tool needs to be systematically evaluated. Such a methodical approach should allow transfer into advanced coating design with more reproducible and enhanced properties. Several other related issues are contained in Tables 1.18 and 1.19.

Surface technologists are central to the creation and development of a

Raw materials
- **Chemistry**
- **Manufacturing process**
- **Powder characteristics etc**

Part surface preparation
- **Dry/wet grit blasting**
- **Shot peening**
- **Machined stripped surface etc**

Coating application
- **Conventional air plasma**
- **High velocity oxyfuel**
- **Gator-Gard etc**

Post-coating treatment
- **Heat treatment**
- **Sealing**
- **HIPing etc**

Finishing
- **Machining**
- **Grinding/lapping**
- **Slurry abrasive finishing etc**

Quality assurance
- **Metallography**
- **Bond strength testing**
- **Hardness etc**

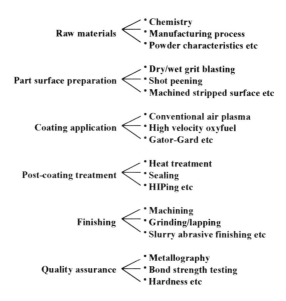

Figure 1.8 Steps involved in the production of thermally sprayed coatings [8].

commercial coating through a suitable process technology – see for instance Fig. 1.9 in the case of thermal spraying. These engineers and scientists will need to be sufficiently aware of current research in the specified coatings area, and recognize the significance of the properties and characteristics of the coatings in relation to anticipated performance demanded by the end-user. Clearly it is especially important in the creation of a novel coating, that they also understand fundamental aspects of the design of the coating equipment, and especially how the process

Table 1.18 Future surface engineering activities [5,7]

1. Surface engineering of non-ferrous metals
2. Surface engineering of polymers and composites
3. Surface engineering of ceramics
4. Mathematical modelling of surface engineered components
5. Surface engineering in material manufacture
6. Statistical process control in surface engineering
7. Non-destructive evaluation of surface engineered components
8. Duplex or hybrid surface engineering technologies and design, e.g.:
   * laser treatment of thermal and plasma spray coatings
   * ion beam mixing and ion-assisted coatings
   * hot isostatic pressing of overlay coatings
   * thermochemical treatment of pre-carburised steels
   * thermochemical treatment of pre-laser hardened steels
   * CVD treatment of pre-carburised steels
   * PVD treatment of pre-nitrided steels
   * ion implantation of pre-nitrided steels

Table 1.19 Technological and ecological comparisons of different surface technologies for tribological applications [10]

	Pollution	Distortion temperature	Adhesion
Nitriding			
Salt bath	-	-	++
Gas	+	-	++
Plasma	++	+	++
Ion implantation	++	++	++
Coating			
Galvano-techniques	-	++	-
CVD	+	--	++
PVD	++	++	-
PACVD	+	+	+

variables may be exploited and controlled to permit the repeatable deposition of material(s) with appropriate characteristics/properties and therefore performance [11].

The 1990s have seen an increasing concern regarding quality assurance and control issues associated with advanced surface engineering technologies – certain issues of which are illustrated in Fig. 1.10. Quality assurance is central to surface engineering as Strafford [11] has stressed in his definition of quality, viz.: "the production of a given coating or surface treatment by a suitable process or hybrid

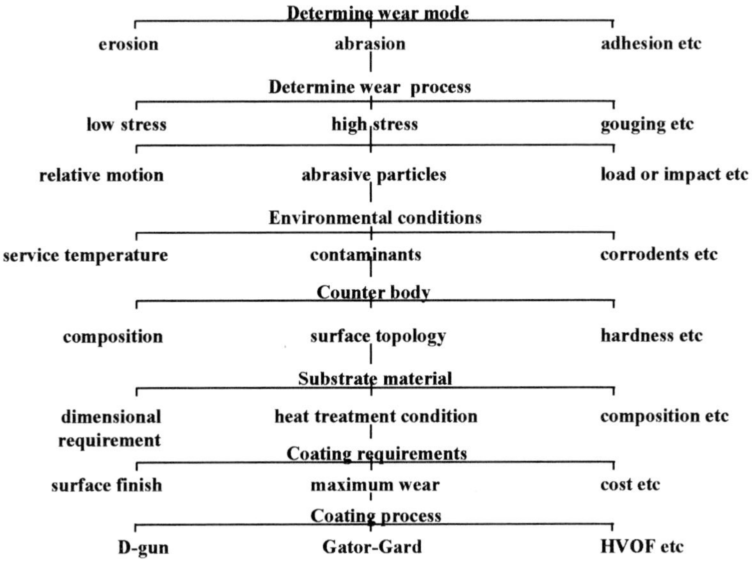

Figure 1.9 Selection of thermal spray coatings and processes [8].

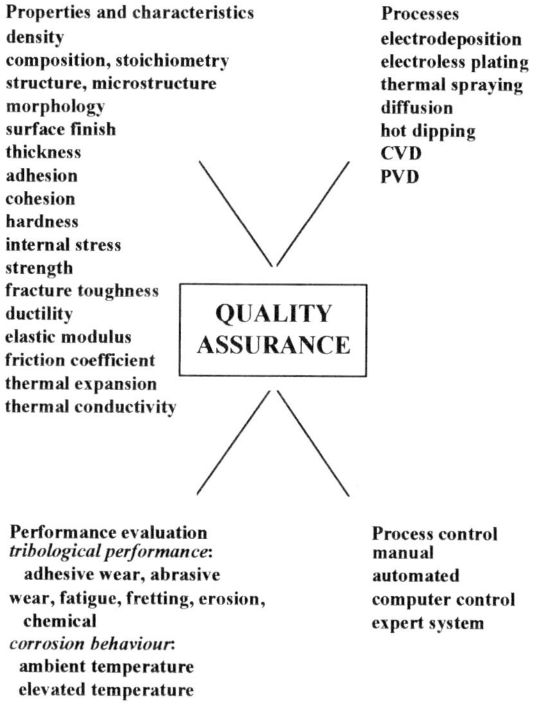

Figure 1.10 Interactive parameters involved in quality assured coatings [19].

technology which reliably imparts the desired measurable performance". Interest persists in the development of expert systems to design and select coating/treatment systems and control the process technology.

References

1. Johnston R, Chambers S, Harland C, Harrison A and Slack H "Cases in Operations Management" Published by Pitman Publishing, 1995.
2. Strafford K N, Datta P K and Gray J S (eds.) "Surface Engineering Practice: Processes, Fundamentals and Applications in Corrosion and Wear" Published by Ellis Horwood, 1990.
3. Datta P K and Gray J S (eds.) "Surface Engineering" ,Vol. I Fundamentals of Coatings, Vol. II Engineering Applications, Vol. III Process Technology and Surface Analysis
   Published by the Royal Society of Chemistry, 1993.
4. Burnell-Gray J S and Datta P K (eds.) "Surface Engineering" To be published by the Royal Society of Chemistry, 1996.
5. Sankaran V, "Surface Engineering – A Consultancy Report" Advances in Materials Technology: Monitor, Issue 24/25 February 1992.

6. Strafford K N and Subramanian S, *J. Materials Processing Technology*, **53**, p393–403, 1995.
7. Bell T, J. Phys D: *Appl Phys*, **25**, 1A, A297-A306, 1992.
8. Sahoo P, *Powder Metallurgy International*, **25**, 2, p73–78, 1993.
9. Hemsley D, *Engineering*, **235**, 9, p25-27, 1994.
10. Grün R, *Surface and Coatings Technology*, **60**, (1-3), p613–618, 1993.
11. Correspondence with K N Strafford, University of South Australia.
12. Datta P K and Burnell-Gray J S, Internal report for the Surface Engineering Research Group, University of Northumbria at Newcastle, 1988–1996.
13. Cowan R S and Winer W O, *J. Phys D: Appl Phys*, **25**, 1A, A285–A291, 1992.
14. Wright G and Strydom L R, *South African Mech Eng*, **38**, p467–471, 1988.
15. Sioshansi P, *Thin Solid Films*, **118**, p61–71, 1984.
16. Bailey D, Chandler P, Raymond P and Nicoll A R, *Mater Design*, **9**, p330–333, 1988.
17. Hoff I H, *Welding and Metal Fabrication*, **63**, 7, p266–269, 1995.
18. Rickerby D S, and Matthews A, "Advanced Surface Coatings: A Handbook of Surface Engineering" Published by Blackie, 1991.
19. Subramanian C, Strafford K N, Wilks T P, Ward L P and McMillan W, *Surface and Coatings Technology*, **62**, p529, 1993.

Chapter 2

Overview of surface engineering technologies

A Matthews – University of Hull

2.1 Background

As indicated in Chapter 1 there are many ways of categorising and classifying the surface coating and treatment methods currently available. One of the first categorisations was that proposed by James [1], which is shown in Fig. 2.1. This highlights the enormous range of techniques available, many of which can be further sub-divided into dozens of derivative methods. When one also adds the further variable of the hundreds of possible substrate choices, the task of selecting the appropriate technique for a particular application becomes extremely daunting. Nevertheless, such is the importance of surface engineering to product quality and reliability, it is necessary that design engineers and others involved in optimising performance and preventing failure have a reasonable knowledge of the main features of the coating and treatment methods at their disposal. Here we shall overview some of these; as a background to the applications-related aspects covered in the rest of the book. Table 2.1 summarizes some of the main characteristics of the main generic groups of process types.

2.2 Thermal treatments

The thermal and thermo-chemical treatments were probably the earliest methods used to improve the surface properties of metals. The usual aim is to change the type and size of the grain structure. Although they can be carried out on non-ferrous materials, they are most often applicable to steels. A typical example is the quench hardening process. In order to understand how this works, it is necessary to know something of the metallurgy of carbon and alloy steels.

Pure iron changes its crystal structure from body-centred cubic to face-centred cubic as it is heated beyond 910°C. The lower temperature phase is known as *α-iron* or *ferrite*. It is soft and ductile. The interatomic spaces are small, and thus the interstitial solubility of carbon in ferrite is low (i.e. ~ 0.02%). There is a higher temperature phase (910–1400°C) called *γ-iron* or *austenite*. In this

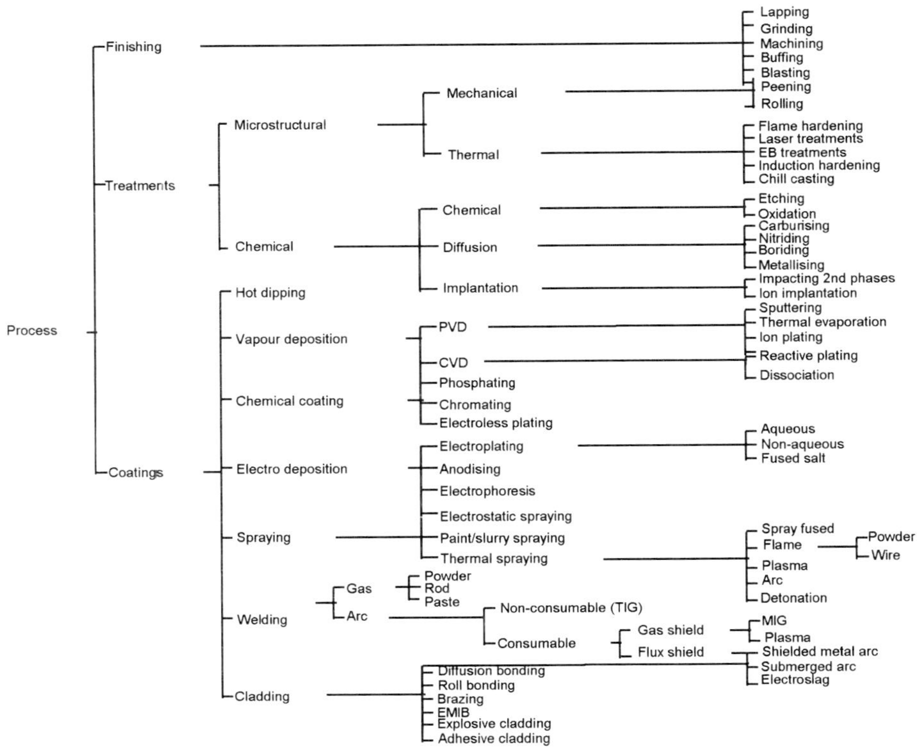

Figure 2.1 Process groups in engineering.

Table 2.1 Comparative characteristics of some of the main coating methods

	Gaseous state process				
	PVD	PAPVD	CVD	PACVD	Ion implantation
Deposition rate (kg/hr)	Up to 0.5 per source	Up to 0.2	Up to 0.1	Up to 0.5	
Component size	Limited by chamber size				
Substrate material	Wide choice	Wide choice	Limited by deposition temperature	Some restrictions	Some restrictions
Pre-treatment	Mechanical/ chemical plus ion bombardment	Mechanical/ chemical plus ion bombardment	Mechanical/ chemical	Mechanical/ chemical plus ion bombardment	Chemical plus ion bombardment
Post-treatment	None	None	Substrate stress relief/ mechanical properties	None	None
Control of deposit thickness	Good	Good	Fair/good	Fair/good	Good
Uniformity of coating	Good	Good	Very good	Good	Line of sight
Bonding mechanism	Atomic	Atomic plus diffusion	Atomic	Atomic plus diffusion	Integral
Distortion of substrate	Low	Low	Can be high	Low/moderate	Low

	Solution process		Molten or semi-molten state process		
	Sol-gel	Electro-plating	Laser	Thermal spraying	Welding
Deposition rate (kg/hr)	0.1–0.5	0.1–0.5	0.1–1.0	0.1–1.0	3.0–50
Component size	Limited by solution bath		May be limited by chamber size		
Substrate material	Wide choice	Some restrictions	Wide choice	Wide choice	Mostly steels
Pre–treatment	Grit blast and/or chemical clean	Chemical cleaning and etching	Mechanical and chemical cleaning		
Post–treatment	High temperature calcine	Non-thermal treatment	Non-substrate stress relief		None
Control of deposit thickness	Fair/good	Fair/good	Fair/ good	Manual–variable Automated–good	Poor
Uniformity of coating	Fair/good	Fair/good	Fair	Variable	Variable
Bonding mechanism	Surface forces		Mechanical/chemical		Metallurgical
Distortion of substrate	Low	Low	Low/ moderate	Low/ moderate	Can be high

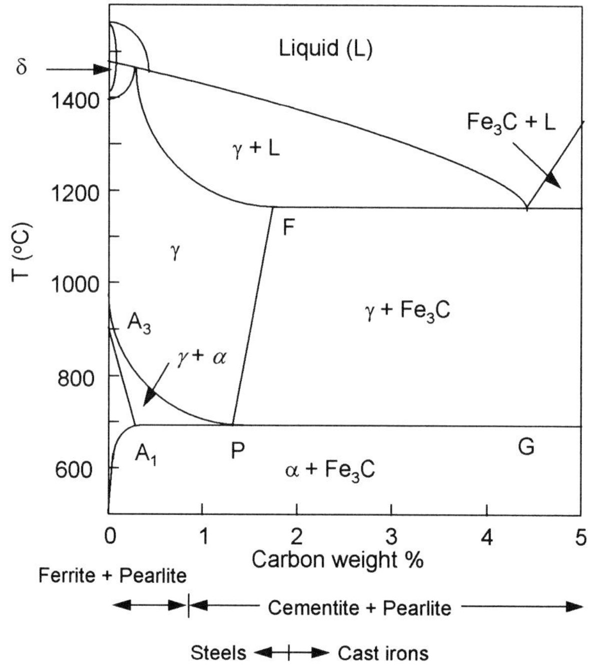

The ferrite solubility line, A_1P, denotes the commencement of precipitation from austenite. The cementite solubility line, FP, indicates the primary deposition of cementite from austenite.The pearlite line, A_1PG, indicates the formation.

Figure 2.2 Part of the iron–carbon equilibrium diagram.

temperature range also it is soft and ductile – making it possible to be forged and rolled. The solubility of carbon in austenite is higher than for ferrite, but still less than about 2% by weight (the '*solubility limit*'). In iron–carbon alloys, carbon concentrations in excess of the solubility limit can lead to the formation of a secondary iron carbide phase, known as cementite Fe_3C. This has a crystal lattice in which iron and carbon are in the ratio 3 to 1. It contains 6.67% carbon by weight. Cementite is very hard and brittle, however its presence with ferrite in steel considerably increases the strength if the structure and distribution is closely controlled.

When austenite of the eutectoid composition (0.8% carbon) is cooled to below the eutectoid temperature, ferrite and cementite are simultaneously formed, with a lamellar structure (alternate layers of ferrite and carbide). This composite phase is known as *pearlite*. There are many microstructural forms which carbon can take up within steels when they are cooled, depending for example on the carbon content and the rate of cooling.

One important form is known as *martensite*. This has a fine needle-like structure, is very hard, and is formed by shear transformation when the steel is rapidly quenched from high temperature. The actual phases which result can be described by use of a TTT (*time temperature transformation*) diagram.

It may be possible to improve the properties of the quenched product by further heat treatments, such as *annealing* and *tempering*. Also, additions of various alloying elements to carbon steel can significantly change its heat treatment properties. With the exception of cobalt, they all retard the onset of transformations. Thus an austenitic steel is one whose alloy content has prevented the transformation from austenite upon cooling to room temperature.

In the context of *thermal treatments* applied specifically to improve the surface hardness, there are a number of heating methods available – including *induction*, *resistance*, *flame*, *laser* and *electron beam heating*. Of these, *induction heating* is the most widespread. Typically a medium carbon (0.3–0.6% C) content steel is used, and the depth of the hard case produced is usually 1–10 mm.

Like thermal treatments, *thermochemical treatments* rely on *modifying* the existing material rather than adding a coating. For example, carbon or nitrogen can be diffused into the surface to produce a hard martensitic layer, or a fine dispersion of nitrides and carbides in combination with alloy elements (e.g. Al,

Table 2.2 Some thermochemical treatments based on carbon and nitrogen

Process	**Description**	**Process**	**Typical treatment temp. (°C)**	**Typical case depth (mm)**	**Typical surface hardness (H_v)**
Carburising	A process in which a steel surface is enriched with carbon, at a temperature above the ferrite/austenite transformation. On subsequent quenching, an essentially martensitic case is formed.	Solid Liquid Gaseous Plasma	850–950	0.25–4.0	700–900*
Carbonitriding	Similar to carburising, but involving nitrogen as well as carbon enrichment.	Liquid Gaseous Plasma	700–900	0.05–0.75	600–850*
Nitrocarburising	A process in which a steel or cast iron surface is enriched with nitrogen, carbon and possibly sulphur at a temperature below the ferrite/austenite transformation.	Liquid Gaseous Plasma	570	0.02 max† 1.0 max‡	500–650†
Nitriding	A process in which a steel surface is enriched with nitrogen, at a temperature below the ferrite/austenite transformation.	Gaseous Liquid Plasma	500–525	0.4–0.6	800–1050

* Depending on temperature treatment (upper figure represents typical as-quenched hardness).

† Thickness and microhardness of compound layer on mild steel. Values are dependent on alloy content of material.

‡ Total depth of diffusion.

Cr, Mo, V). There are various source media which can be used for the nitrogen or carbon – such as solid (normally called '*pack*' processing), liquid, gas or plasma (ionized gas). The main thermochemical processes are summarized in Table 2.2 [2].

The *carburising* process is usually applied to steels with a low initial carbon content (< 0.45% C). Special steels have been developed for nitriding, which contain suitable nitriding elements. A popular nitriding steel for many years was EN 41, nowadays '*hot working*' tool steels such as AISI H13 (a 5% Cr steel) are often *nitrided*. There are several proprietary *carbonitriding* and *nitrocarburising* processes (e.g. '*Tufftride*' or '*Sulfinuz*') which are carried out at comparatively low temperature (e.g. 570°C) and can be produced on a wide range of steels including low carbon steels. With many of these types of processes a porous compound layer can be produced, which should be removed prior to in-service use – or (as in the '*Nitrotec*' process) it can be combined with a sealant to enhance the corrosion resistance [2,3].

In addition to these thermal techniques, there are also *mechanical methods* of surface modification, such as *peening*; these are often intended to produce beneficial residual stresses in the surface, in order to enhance fatigue life for example. Both the thermal and mechanical surface modification methods have been around for many years, and are thus already well accepted by industry.

2.3 Coating methods

The coating methods have perhaps been less well accepted by industry; this is the case in particular for the newer or advanced coating techniques, and we shall thus give these special attention, since they can be enabling technologies which make possible technical developments which were previously unachievable.
A useful approach to the generic categorisation of coatings is given in ref. [4], which is further modified in ref. [5], as shown in Fig. 2.3. This divides the methods into *gaseous state*, *solution state* and *molten* or *semi-molten state* processes. Typical thickness and processing temperature ranges for some of the main methods are shown in Fig. 2.4 (a) and (b)

2.3.1 Gaseous state processes

The gaseous state processes include *physical vapour deposition* (PVD) and *chemical vapour deposition* (CVD), as well as various plasma- or ion-based variants of these methods.

The CVD process involves the reaction of gaseous reagents within a vacuum vessel, so that they form a deposit on the heated sample surface. The technique is used primarily for the deposition of ceramics such as titanium carbide (TiC), titanium nitride (TiN) and alumina ($A1_20_3$). Since the deposition is typically carried out at a temperature above 900°C, the process has only found widespread usage

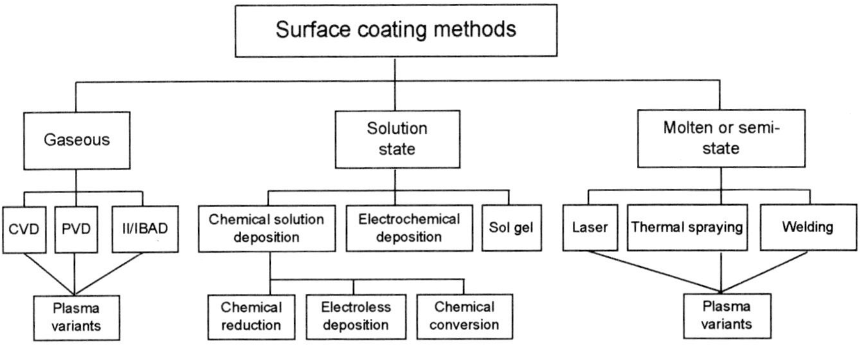

Figure 2.3 A general classification of surface engineering techniques.

on materials which can withstand such a temperature without softening or appreciable distortion. Such a material is cemented carbide (the WC–Co composite used widely for lathe cutting tools).

The PVD process, like CVD, can be used for the deposition of pure ceramic coatings (as well as metals and alloys). However it differs from CVD in that at least one of the constituents is physically evaporated from solid within the vacuum chamber. In the most advanced processes the sample to be coated is made the cathode in a glow discharge of the evaporated metal and gas atomic species. Thus, for example, titanium can be evaporated in nitrogen to produce titanium nitride. The benefit of the ionisation which results in the glow discharge is that the positively charged depositing species are accelerated to the sample surface and arrive with high energy – producing a dense, well adhered deposit. Also,

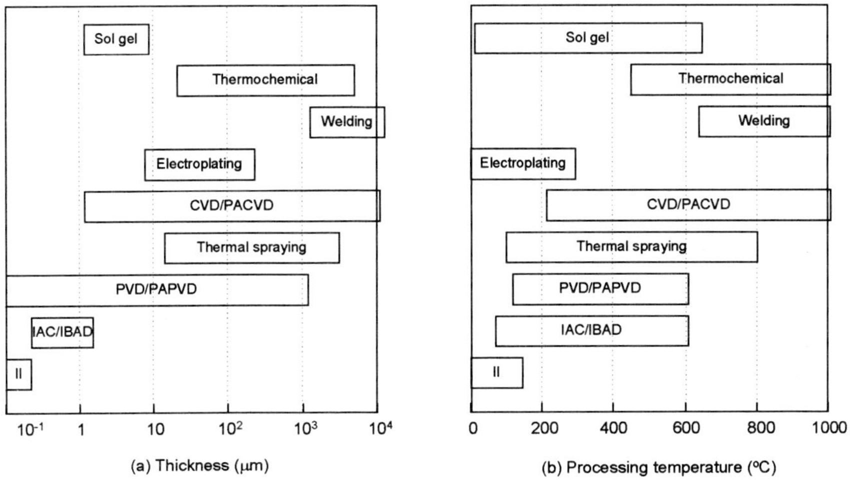

Figure 2.4 Typical ranges for (a) depths of surface modification and thickness of coatings, and (b) processing temperatures for surface technologies.

since the energy is imparted at the surface, where it is needed, the process can be carried out at comparatively low bulk substrate temperatures (i.e. $< 500°C$). Thus materials such as hardened high speed steel and hot working tool steels can be coated with pure ceramic films without being softened. There are various ways of producing the metal evaporant – such as electron beam guns, 'sputter' sources and arc sources. The PVD process is highly flexible, permitting the deposition of virtually any material onto any other material, and is discussed further in this chapter.

2.3.2 Solution state processes

The solution state processes include *electroplating* and *conversion* processes such as *anodising*. So-called '*hard chrome*' electroplating is probably the most widespread wear resistant coating in this category. This differs from conventional decorative chromium electroplate because it is thicker (i.e. 10–500 μm cf 1 mm) and usually deposited onto a hard underlying material. The typical hardness for the coating is 600–750 H_v. There are various pre-treatment and post-treatment procedures recommended for hard chrome, and indeed other electroplates, as discussed in [2]. One effect to be aware of is *hydrogen embrittlement* – which is caused by the adsorption of hydrogen during deposition from aqueous solutions. This can be detrimental to the fatigue life, but its effects can be reduced by heat treatment procedures.

Electroplated hard chrome tends to form cracks during deposition which can be beneficial for lubricated applications as the cracks can retain the lubricant. One of the reasons for the good friction and wear performance of the electroplated hard chrome is a thin oxide on its surface. However, if the contact pressure is too high during operation, this protective layer breaks down – resulting in rapid wear.

Other materials which are commonly electroplated are nickel alloys, copper, cobalt alloys and iron alloys, as well as various other soft metals and alloys for applications such as electrical contacts and plain bearings.

A recent innovation has been the development of electroplating processes in which a fine powder of a ceramic material (such as alumina) is present within the electrolyte. By agitating this during deposition, it is possible to produce a coating which comprises a ceramic within a metal matrix.

Another electrochemical process mentioned above is *anodising*. This usually involves the use of a sulphuric acid solution, in which an aluminium component to be treated is made an anode. This leads to the conversion of the surface to hard aluminium oxide to a depth of 25–150 μm. Post-treatment procedures are available to improve the corrosion resistance, these can however be detrimental to the wear life. It is probable that anodising will be used increasingly in the future – as aluminium will be more widely specified due to its light weight, particularly in automotive and aerospace applications. Anodising procedures also exist for zinc and titanium.

Electroless nickel coating is an increasingly important electrochemical technology, which utilizes a process known as *auto-catalytic chemical reduction*. Metal cations in solution are reduced to the metal by a chemical agent, with the surface of the workpiece catalysing the reaction. The process can coat complex shapes with a hard (approx. 550–950 H_v) uniform layer thickness (up to about 5 mm). Also, as it requires no power or electrical contacts it can be applied to non-conductors with suitable preparation. A drawback of the technique is the high cost and relative inefficiency of the reducing agents, and the unavoidable solution losses. In electroless nickel coating, compounds containing phosphorus are usually used as reducing agents – as a result the final deposit contains phosphorus. This has been shown to give it excellent tribological (friction and wear) properties when heat-treated – as discussed in ref. [2]. Recently nickel/powder composites have become available. The most common powders used are diamond, silicon carbide, alumina (in platelets) and PTFE. Due to the increasing importance of electroless nickel coatings they are discussed in some detail in Chapter 3.

2.3.3 Molten or semi-molten state processes

The molten or semi-molten state processes include the "*hard-facing*" methods, such as thermal and plasma spraying. A very wide range of materials can be deposited by these techniques, including aluminium alloys, copper, iron, molybdenum, nickel, stainless steel and tin. The thickness produced can be from a few microns to several millimetres, and this represents an economical way of applying wear and corrosion resistant coatings. However, the porosity can often be greater than acceptable, and several process developments, such as *detonation (D)-gun* and *high velocity oxy fuel* (HVOF) thermal spraying, aim to improve this characteristic.

Various *welding* processes can also be used to deposit a range of metals and metal/ceramic composites. The two most widely used methods are the *oxyacetylene torch*, in conjunction with a welding rod composed of the metal to be deposited. The second most common method is to use the arc welding technique, with the welding electrode as the '*facing*' metal.

A semi-molten state process which has recently attracted increased attention is the *friction surfacing* technique [6,7]. This involves rotating the coating material, usually in rod form, at high speed and forcing it against the surface to be coated. Friction heating results in the rod material smearing over the surface. This can produce a well-bonded coating with a dense structure.

Wear resistant coating materials used in molten or semi-molten state processes include carbides of chromium, tungsten or boron dispersed in a matrix of iron, cobalt or nickel. Those used to resist impact generally have austenitic structures that can work harden, e.g. austenitic manganese steel. This as-deposited is relatively soft (approx. 170 H_v) but hardens to about 550 H_v when subjected to impact conditions. This material gives an example of how the processing route can affect the coating properties. The austenitic structure of manganese steel is normally produced by a water quench from about 980°C. The welded coating alloys are often formulated to ensure that the austenitic structure is obtained on air cooling from the welding temperature. Aspects such as this require specialist advice from the suppliers of deposition equipment and welding consumables. One material which is particularly widely used is the cobalt-based alloy *Stellite*. Various grades of this material are available, to meet specific operating needs. Publications such as [2] provide further details on the optimum properties for this and other coatings. In this single chapter it will not be possible to cover all of the available coating methods in great detail. In the coming sections we shall thus give greater prominence to several emerging techniques, beginning with the gaseous (or vapour) state deposition routes and their ion or plasma-based variants. Following this will be information on electroless nickel coatings, an increasingly important technology in the solution-state processing sector. Then, information will be given on thermal spraying, representing the molten or semi-molten state process sector.

2.4 Gaseous state processes

2.4.1 Process details

The main techniques in this group are *chemical vapour deposition* (CVD) and *physical vapour deposition* (PVD).

As discussed in [8] and illustrated in Fig. 2.5, there are many variants on the basic CVD process. *Carlsson* discusses in [4] the main features of the conventional thermally activated CVD processes and cites its applications, which include the growth of free-standing shapes of brittle materials, the densification of surfaces by infiltration techniques, and the production of ceramic coatings for cutting tools.

The use of plasma assistance for CVD has been particularly significant in the electronics industry, through various plasma etching and deposition processes. Also *plasma-assisted CVD* is an important technique used for the deposition of diamond and diamond-like-carbon films [9–11].

For wear and corrosion resistance applications in engineering the PVD techniques are in more widespread use than CVD and they will be discussed below.

PVD systems all incorporate a means of evaporating coating material from a solid source under a partial vacuum. The most advanced ones involve the creation of a glow discharge or plasma, and deposits produced by these ionisation-assisted or 'ion plating' processes possess many benefits, as cited below:

- They are well adhered – due to the high arrival energy of coating atoms and the possibility to '*sputter-clean*' the surface by argon ion bombardment prior to deposition.
- They usually require no finish machining, as they replicate the original surface

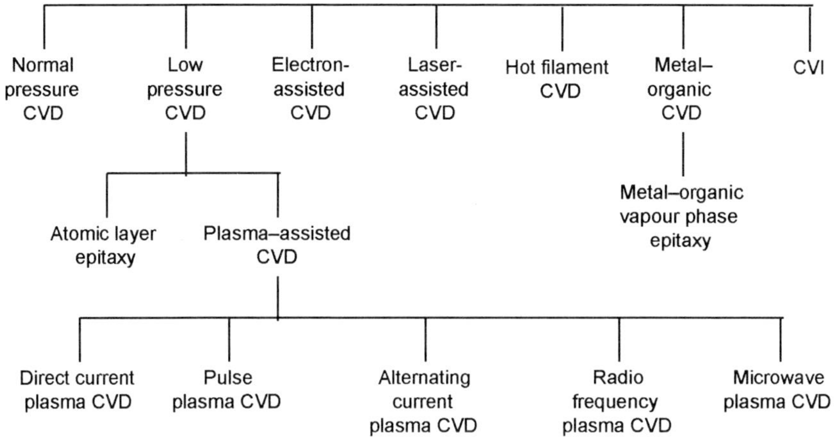

Figure 2.5 Basic CVD methods and ion-based derivatives.

finish.

- Deposition temperatures can be low (e.g. for ceramic deposition < 500°C, cf > 900°C for CVD).
- Thicknesses can be very uniform – due to gas scattering effects and the ability to rotate and reciprocate items relative to the source.
- The coating structure can be controlled.
- A very wide range of coating and substrate materials can be used.
- There are usually no effluents or pollutants produced.
- Deposits have high purity.
- Deposition rates can be finely controlled.
- The '*hydrogen embrittlement*' problems sometimes experienced with electroplating can be avoided.
- The coating systems readily lend themselves to automated batch manufacture.

The main techniques available to evaporate the coating material are:

i. Resistance heated sources. Resistance heating involves holding the evaporant in a holder made from a refractory material such as molybdenum or tungsten, or an intermetallic compound such as titanium diboride/boron nitride. The container is heated by passing a current through it. This method has been used for evaporating low melting point materials such as aluminium, copper, silver and lead [12].

ii. Electron beam guns. Electron beam guns are becoming increasingly popular since there is effectively no limit on the melting point of materials that can be evaporated. The most popular types of gun used are the '*bent beam*' *self accelerated gun* (Fig. 2.6) [13], the *work accelerated gun* (Fig. 2.7) [14], and a variant of the latter, known as the *hollow cathode discharge* (HCD) electron beam gun (Fig. 2.8) [15,16]. The last two systems are very effective in

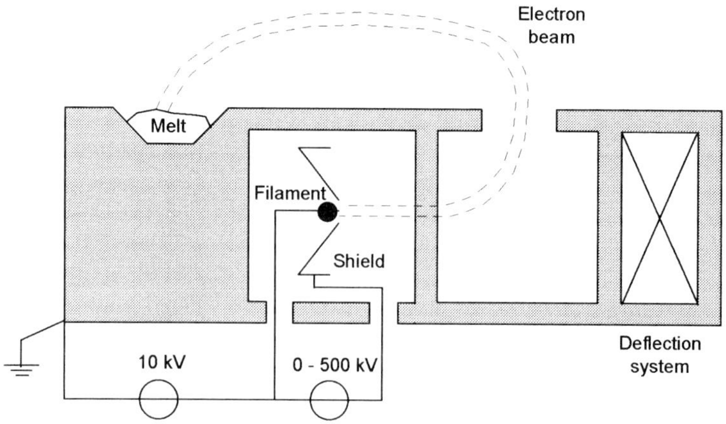

Figure 2.6 A schematic view of a 270° electron beam gun.

increasing the ionisation when used in ion plating systems.

iii. Arc sources. Arc evaporation was used for many years as a means of vaporising carbon – simply by striking an arc between two carbon electrodes. Recently considerable interest has surrounded the use of the arc technique to evaporate metals such as titanium (Fig. 2.9) [17,18]. This layout also enhances the ionisation of the depositing species. A further benefit is that, because no molten pool is formed, evaporators can be fitted in any orientation, including inverted. However, a reported disadvantage for this technique is that droplets are ejected from the source producing '*macroparticles*' in the coatings.

iv. Sputter sources. Sputtering sources operate through the bombardment of the source or 'target' by ions and accelerated neutrals (usually of argon), which in turn leads to the ejection of the source material. A major advantage of this technique is that alloys can be sputter deposited to retain their composition. Two disadvantages are the large power losses that occur with this type of source (95% of the electrical power input can be lost to the cooling water from the target), also the deposition rates can be lower than those achieved

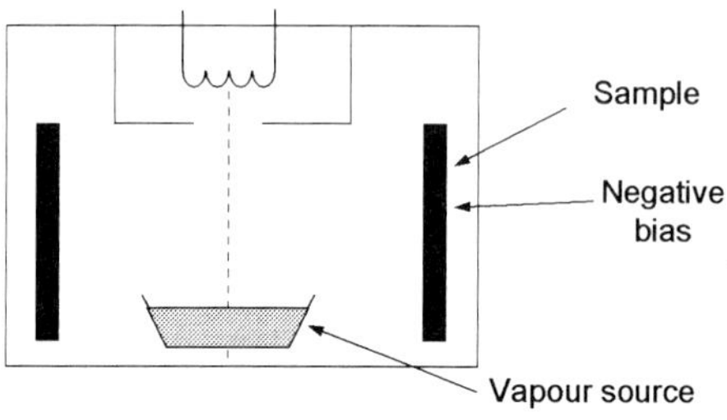

Figure 2.7 The work accelerated EB gun system.

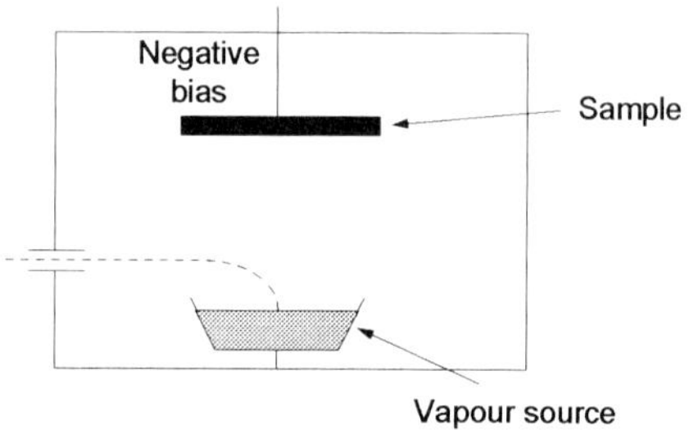

Figure 2.8 A hollow cathode discharge gun system.

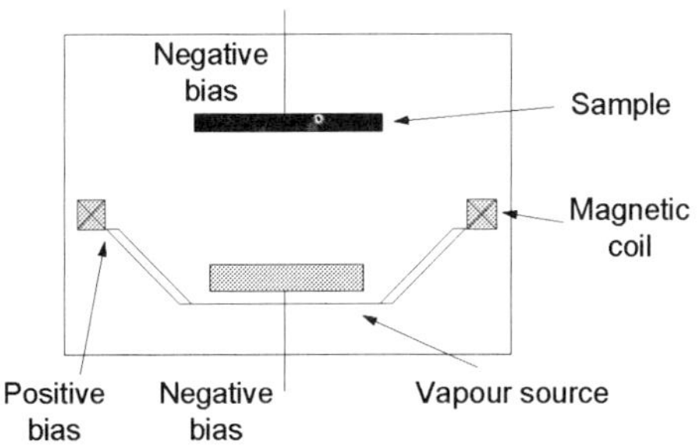

Figure 2.9 An arc source system layout.

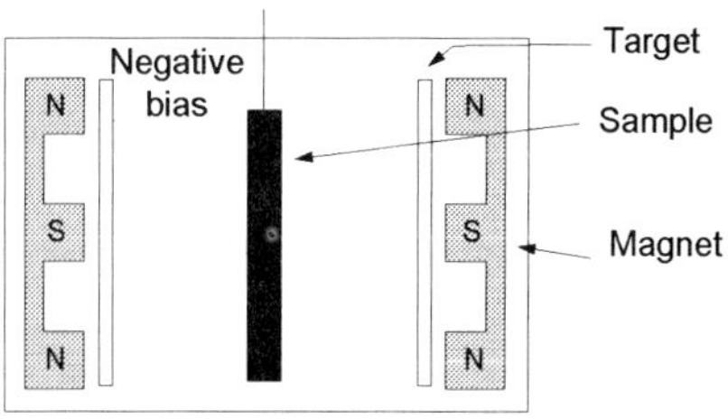

Figure 2.10 A mirrored magnetron system.

by other methods. The latter problem can be overcome by a number of means. Firstly the surface area of the target (or targets) can be made much greater than that of the item to be coated. Secondly, magnetic confinement techniques can be used, to increase and confine the amount of ion and neutral bombardment. Fig. 2.10 shows one system incorporating both these features [19]. Recently this arrangement has been improved by "*unbalancing*" the magnetrons and configuring them so that the magnetic poles are opposed. These two steps have the effect of releasing some of the electrons and ions confined near to the magnetron, so that they can be concentrated near to the samples. Figures 2.11 and 2.12 show how the *mirrored* and *opposed* (or *closed*) field systems compare. References [20] and [21] give more details.

The methods cited above cover only a small number of the many variants of PVD which exist. These are discussed further in [8] and illustrated in Fig. 2.13.

The origins of PVD can be traced back to the nineteenth century, and even the "*advanced*" PVD technique which we now call *ion plating* [22] was first patented 50 years ago [23]. It is only in the last decade however that the method has found wide industrial usage. The reason for this is that during this time techniques for the deposition of ceramics at low temperature have been developed and brought

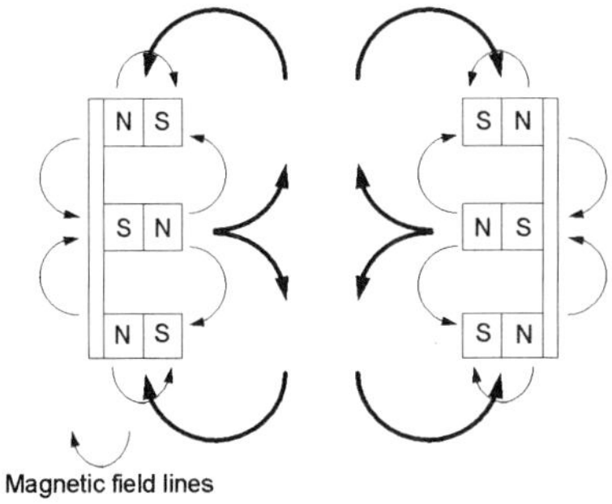

Figure 2.11 The basic form of the magnetic field in a mirrored configuration.

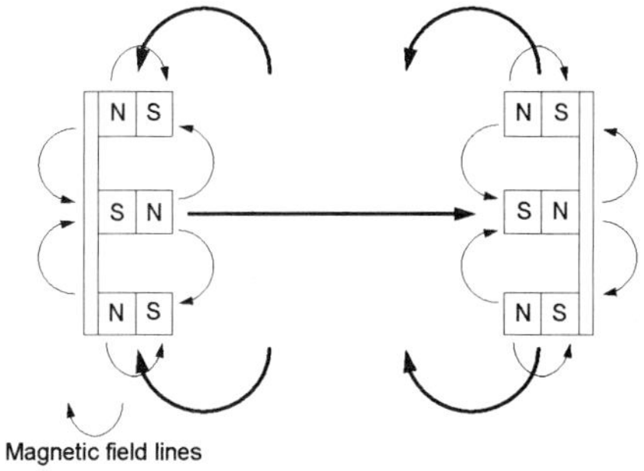

Figure 2.12 The basic form of the magnetic field in a confined (closed field) configuration.

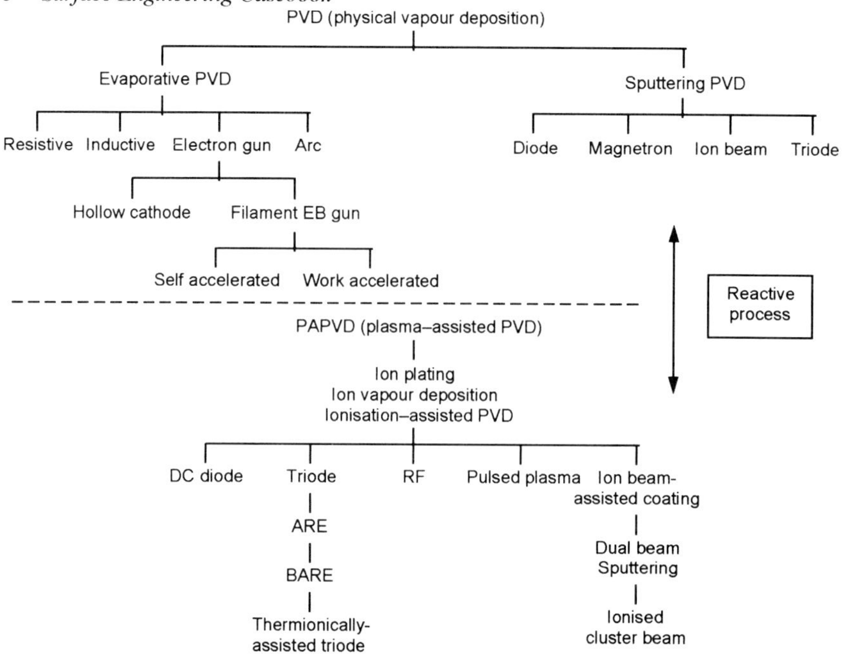

Figure 2.13 PVD processes and derivatives.

to commercial maturity. The particular development which has made this possible has been the production of ionisation-enhancing systems, such as the vapour sources mentioned above. Since it is the ceramic coatings that have had the greatest impact, this section will concentrate on these films.

A typical *process cycle* comprises the following stages:

- Depending on the substrate and coating materials, it is usually necessary to put the components through a thorough pre-cleaning procedure. For wear resistant ceramic coatings on alloy steels, for example, multi-stage cleaning processes are used, incorporating techniques such as ultrasonic fluorocarbon solvent cleaning, acid/chemical etching, ultrasonic water/detergent cleaning and freon drying/dewatering systems. Several of these stages may be repeated, and they may be augmented by additional process stages, such as for rinsing and '*vapour*' washing or wet/dry abrasive cleaning. The technology of cleaning is thus an important part of PVD processing. After cleaning, samples are loaded into the vacuum chamber, which is then pumped down to a pressure below 10^{-3} Pa. The pumping speed is then reduced and the chamber is backfilled with argon to a pressure of about 1 Pa. A negative bias voltage of several kV (DC) is then applied to the component to be coated. This results in the initiation of a glow discharge and the bombardment of the component by argon ions and neutrals. This *sputter-cleaning* stage is continued for 20–30 minutes to remove loosely bonded atoms, and to drive off other contaminants – such as surface hydrocarbons. Also, if necessary, thin oxide layers can be removed. In the case of the arc deposition systems, this cleaning stage is achieved with titanium ions – by energising the source and ensuring no net deposition on the component.

- The next stage of the process (again depending on the system and the coating material) involves steps to increase the substrate temperature. In the case of the *work accelerated gun* this can be achieved by electron bombardment (biasing the component positively). The technique used in some laboratories and processing companies is to utilize a '*thermionic triode*' ion plating system (Fig. 2.14), which allows the ionisation in the chamber to be increased independently of the vapour source. This facilitates intense ion bombardment of the sample surface, and predictable component heating. Normally the deposition stage for ceramic coating is carried out at reduced pressure and voltage. Typically a reactive gas such as nitrogen will be reacted with a vaporized metal such as titanium, to produce, in that case, titanium nitride.

A number of commercial systems are now available. Much research effort is currently being devoted to improve these, for example to enhance our understanding of the complex plasma processes occurring in ionisation-assisted PVD, and in monitoring and controlling the systems. Techniques such as *optical emission spectroscopy* are being used increasingly to ensure repeatability in the coatings produced.

2.4.2 Applications for PVD

Cutting tools

The gaseous or vapour state processes can be used in a wide variety of applications. A considerable amount of data is available on the applications of titanium nitride, e.g. [24–31]. The main application has been for cutting tools, this may be in part due to the fact that the industry had already become acquainted with ceramic coatings such as titanium nitride, produced by CVD, for tools such as turning inserts.

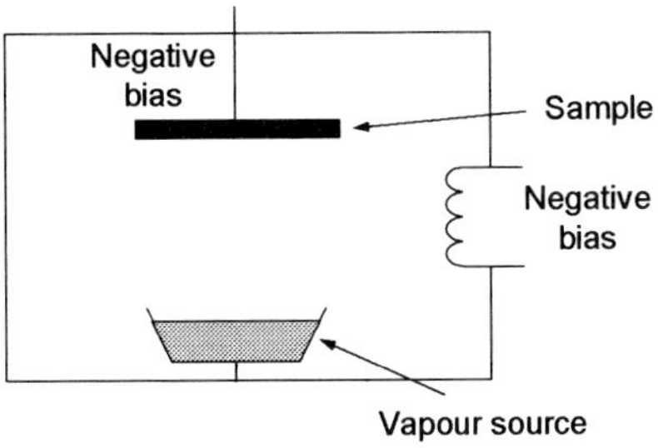

Figure 2.14 A thermionically–assisted triode system.

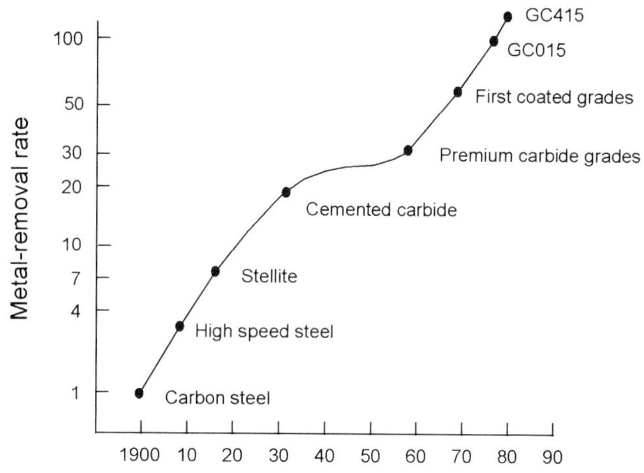

Figure 2.15 The chronological influence of new tool materials and coatings on metal removal rates.

The impact which coatings can make on tool performance has been well illustrated by *Sjorstrand* [25], in the case of CVD (Fig. 2.15). In that process, multi-layered deposits are already in widespread use. Although other ceramics offer considerable potential to improve upon titanium nitride the results being achieved with this material are quite remarkable. This is particularly evident on drills and gear cutters; these will therefore be discussed below.

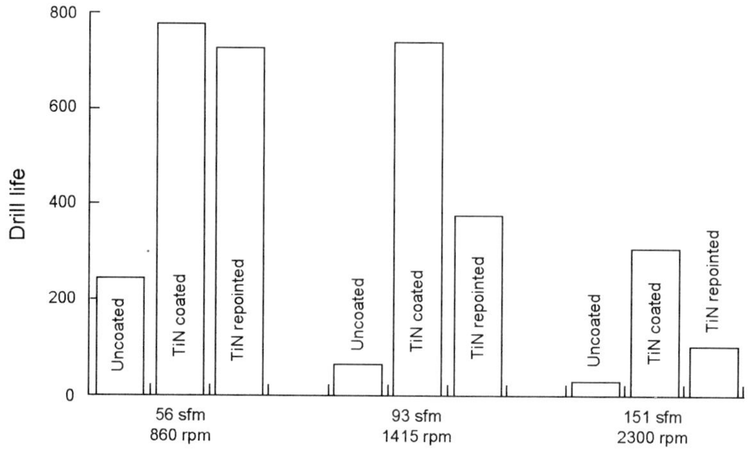

Figure 2.16 The influence of cutting speed and regrinding on coated and uncoated drill life.

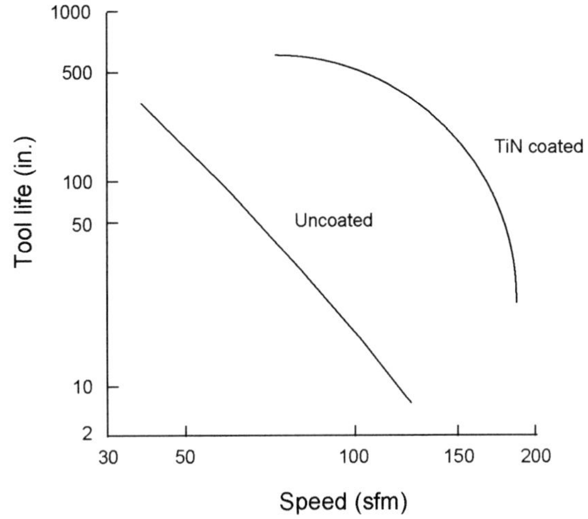

Figure 2.17 The effects of speed on coated and uncoated drill life.

Drills

There are now many published test results for coated drills, with several manufacturers offering their own products. Figures 2.16, 2.17, 2.18 and 2.19 [26] show a series of early graphs which presented some important comparative information on coated and uncoated drills. First of all it can be seen that the improved performance of the coated drills applies across the whole speed range, though when resharpened they maintain their advantage better at lower speeds.

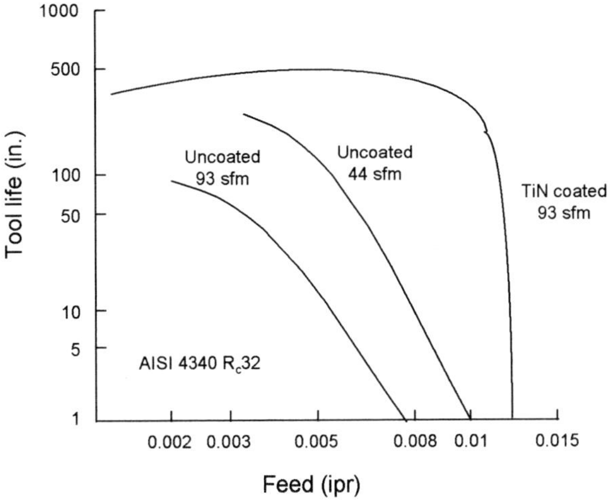

Figure 2.18 The effects of feed rate on coated and uncoated drill life.

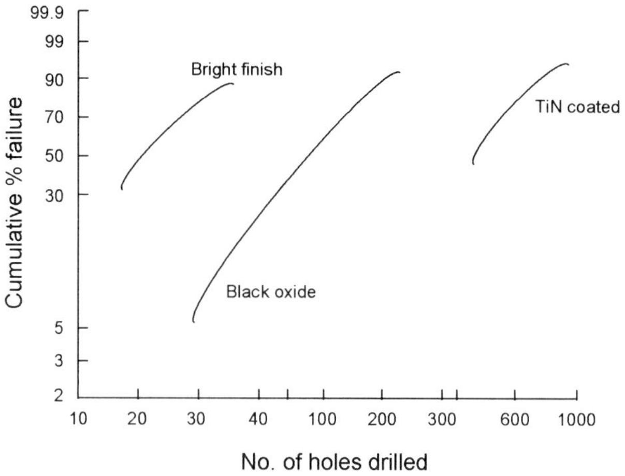

Figure 2.19 The spread of results obtained in drill testing.

Figure 2.16 shows that a coated drill can achieve longer life even when operated at a higher speed than an uncoated one. Figure 2.18 indicates that, for a certain range of feeds, the life actually increases as the feed rate is increased. With drill testing there is always some scatter in results, and this is demonstrated in Fig. 2.19.

Sproul and Rothstein [27] have made an extensive comparative study of sputtered and reactively evaporated titanium nitride coatings on twist drills. They found that the sputtered deposits did not achieve the highest level of life increase given by arc and evaporative processes, but the spread of results on these early non-sputter deposited coatings was very large. For example in unlubricated tests uncoated M7 HSS drills had an average life of 4 holes, cut in 4340 steel (235 BHN) 6.35mm dia. by 1.5cm deep. The sputter coated drills cut 452 holes on average, with a standard deviation of 96. Drills coated by PVD technique 'A' gave an average life of 530 holes, but the standard deviation was 407. PVD technique 'B' gave 398 holes, with a standard deviation of 209. What is most worrying is that the coatings were from the same coating batch in each case. One assumes that *variability* may be even greater from batch to batch. Naturally, especially for companies employing advanced computer controlled manufacturing techniques (where tool life must be predictable to within a few percent) these types of results are quite alarming, as they cannot assume the maximum life increase in each case. Indeed it is better to use a technique with a lower average improvement, if that improvement is more consistently achieved. Studies on the wear mechanisms of drills and other tools [28, 29] have aided our understanding of how the coating alters the cutting conditions (for example by modifying the *built-up edge* BUE) and subsequent work has achieved much more consistent results by altering the drill geometry. In reference [28] a schematic comparison was presented of *flank wear* development on TiN coated and uncoated tools.

This postulated that a smaller *built-up edge* was formed with the coated tool, and that this allowed the more rapid progression to a lower wear rate than possible with an uncoated tool. This mechanism has now been proved to occur, in a remarkable video produced by Dr E D Doyle, whilst at the *Materials Research Laboratories* in Australia. Dr Doyle is presently at the *Centre for Integrated Manufacture* within *Swinburne Institute of Technology*, and also at the company *Surface Technology*. His video (recorded in an SEM) showed how the BUE on an uncoated tool is much larger, leading to '*undersize*' machining when cutting externally. Naturally (although not designed to simulate drilling) this also explains the tendency to produce *oversize* holes when using uncoated rather than coated drills. Also the improved surface finish produced by coated tools can be seen to arise from the reduced tendency for the BUE to break away during cutting. A discussion of hole size accuracy and surface finish is included in [29].

Another finding by Sproul was that, when cutting fluids are used with these coatings, the results may not be as one might expect. Although one fluid doubled the number of holes drilled, a high sulphur content fluid actually reduced the average number of holes cut in one test from 150 to 10. The reason for this is not clear. There was an absence of *fatty ester extreme pressure additive* in the fluid which could be important. Work is continuing to try to explain the results. In the absence of any proven theory manufacturers must clearly be careful in selecting cutting fluids for use with these coatings.

One of the leaders in the application of TiN technology to drills has been the company *SKF* and *Dormer Tools*. They developed a drill geometry (designated ADX) which has a convex lip shape to give more '*bulk*' at the outer edge, and thereby more load support where needed. There is also more room for swarf removal to be accommodated when the drill is used at the high feeds and speeds made possible by coating. Using this drill the company has demonstrated some remarkable results, including:

- increases in peripheral speeds by 1.5 to 2 times,
- increases in feed rates by 25–40%,
- often no need for spot drilling,
- more consistent hole size and finish,
- improved chip configuration for automatic conveyors, and
- performance levels equivalent to carbide drills, but at 50% of the cost.

Gear hobs

Whilst drills are an important potential market for ceramic coatings, it appears that equal, if not greater, success is being achieved on larger tools, particularly gear cutting hobs. There are several reasons for this. Firstly there is a clearer economic benefit in coating (say) a £100 hob for £30 and making it last 4 to 5 times longer, than achieving the same sort of saving and performance improvement on a £1 twist drill. The latter items have traditionally been 'disposable' and less subject to performance monitoring. Also it seems that the early scatter in results mentioned above was wider on small tools like drills than on larger cutting tools. This has been attributed to several factors, including an increased risk of overheating and softening small diameter, low thermal mass, items. The results of Sproul's tests certainly seem to support this (sputtering being a lower temperature process). Some variation may result from substrate material variability, which influences both coating adhesion and bulk strength. Indeed, it has been demonstrated that the maximum benefit of coatings can be achieved by paying special attention to the substrate material, especially by utilising the latest powder metallurgy high speed steels produced by sintering processes. The best known of these are the ASP grades. *Kloster Speedsteel* of Sweden recommend that the metal carbides within the bulk structure should be finely distributed, and the hardening temperature must not be so high as to cause these carbides to be dissolved.

The *David Brown Gear Company* [30] compared a TiN coated *ASP* grade material hob to an uncoated M2 hob. The tool was for gear cutting and the results can be summarized as follows:

- reduction in machine time 33%
- increase in components per cycle 900%
- reduction in tool wear 20%

The total increase in cost on this hob was 89%. Had the tool been used to its full wear life, the increase in output bought for this additional expense would have been of the order of 1080%. This is the immediate benefit and does not include '*hidden*' advantages associated with reduced set-up costs and down-time savings. These results should not necessarily be considered as the norm, but they do indicate what can be achieved with ionisation-assisted PVD TiN coatings.

Another example in the gear-cutting industry was one company which machined small spur gears in EN 36 material [31]. The blanks were normalized and fully tempered prior to gear cutting. Normalising was performed in order to aid stability in case-hardening, and the tempering to aid machinability. The machining followed the route: turn, gear cut external gear teeth, and then spline the bore. On one batch of blanks, the tempering operation had been omitted. The material UTS was thus 1080 MPa instead of 780 MPa. Turning and gear cutting of the external teeth were performed without any problem. Cutting the six teeth for the splined bore was, however, impossible. Flank wear on the shank gear shaper tool (in ASP23 material at 66 HRC hardness) was so severe that each tool was incapable of completing even one component to the required dimensions. Resoftening the blanks was not practicable as this caused distortion and suface finish degradation. It was therefore decided to try a titanium nitride coating on one of the tools. This produced 19 components between regrinds (the previous best on correctly heat treated blanks had been 6) and the amount removed in regrinding the tool was halved.

Mr W Clark of the *David Brown Gear Company* has pointed out that there may still be a rôle for CVD coatings on certain HSS gear cutting tools. He suggests that this process is applicable to unground form tools which are normally finish machined, coated and then hardened, without further machining apart from grinding bores, location faces and proof diameters. The CVD process is less sensitive to contamination of the substrate and this is said to be a major advantage in the case of unground form tools which are difficult to render chemically clean. It should, however, be pointed out that the dimensional accuracy and surface finish of these tools is generally somewhat inferior to the PVD coated ground product. It is also interesting to note that the *Toshiba Tungalloy Division* has stated that there is an important rôle for PVD coatings in the traditional domain of CVD coating – that of the inserted sintered carbide cutting tool [32]. Where high cutting forces are required (as in milling stainless steel) the PVD coating is said to be less prone to chipping than CVD coated carbide.

Hatschek [26] refers to data from *Barber Coleman* in the USA, who claim a coated hob-life improvement of 8 to 10 times compared with uncoated tools at unchanged feeds and speeds. He contends that the rate of flank wear (claimed to be the dominant wear mode) for the reground cutters is about 1/3 of that for uncoated tools over a wide range of HSS alloys, including M2, M3, M35, T15, REX 76, ASP 23, ASP 30 and ASP 60. As an example of M2, he cites a case history from the *Fellows Company* involving production of a 70 tooth automotive ring gear with a 2.5 inch helical gear-shaper cutter. Coated with TiN it produced 300 pieces per sharpening at a wear per sharpening of 0.25 mm. The uncoated cutter had produced 75 parts between sharpening with 9.3 mm of wear removal per sharpening. Cost per piece was reduced from 20.4 cents to 2.0 cents with the coated tool.

Another trend in the gear hob field, as with drills, is the modification of tool geometries to maximize the benefit of the coating. For example, there is a move towards small diameter single-start hobs, operating at high rotational speeds to give good accuracy and high production rates. Larger rake angles and shorter teeth are also being used, as they are with inserted blade hobs. It is interesting that on turning tools Kankaanpaa and Korhonen [33] report a 200% change in coated lathe-tool life after altering rake and clearance angles by as little as one degree. Such a variation might adversely affect a tool's life if it were uncoated.

As mentioned in reference [24], Hatschek also discusses some observations by coaters and tool manufacturers on cutting geometry. He quotes *Balzers* as saying "*it is probably true that the majority of machines are not capable of operating at the speeds at which the coating operates best, so that changes in tool geometry would be premature*". It is interesting, though, that he also quotes Doall and Onsrud, in the area of end mills and router bits, as saying that changes in geometry (lead angles, flute spacing, etc.) should focus on improved chip removal capabilities to accommodate the increased metal-removal rates of higher feeds and speeds.

A word is also needed about *coating thickness*. Randhawa [34] suggests that tool life is dependent upon the thickness of the applied coating, and that there is a minimum thickness, below which there is a sharp decline in tool life. This is normally taken as equivalent to the peak-to-peak roughness of the tool surface. For hobs and shapers, according to one coating supplier, the minimum thickness is 2 microns, although 3 or 4 micron thickness is preferred. Too thick a coating can be brittle and reduce the extra life by spalling. This is a particular problem in interrupted cutting.

Posti [35] cites the curve shown in Fig. 2.20, taken from [36], which indicates an optimum coating thickness of about 4 microns for a particular hobbing operation.

Given the importance of coating thickness, it is perhaps surprising that coating companies sometimes strongly advocate tool recoating. One would have expected the coating thickness to build up – increasing the risk of spallation and moving the tool dimensions out of tolerance. When questioned about this, one coating company stated that build-up does not appear to occur to the same degree on already-coated surfaces. This is a surprising result, and one which merits further investigation.

Randhawa [34] also indicates that the tool life is strongly influenced by the surface finish of the tool and/or the coating. The tool surface finish is, of course, dictated by the manufacturer. In PVD, the coating generally replicates the tool surface finish, although Randhawa gives measurements of coating roughness showing that coatings applied by unmodified arc deposition are three times as rough as those applied by sputtering. This variation may be small compared to the roughness of the tool itself, and in any case it is claimed that arc deposited coatings can self-polish rapidly when used. Having said that, the presence of the resulting pit left by the removed particle must have some effect on the surface properties.

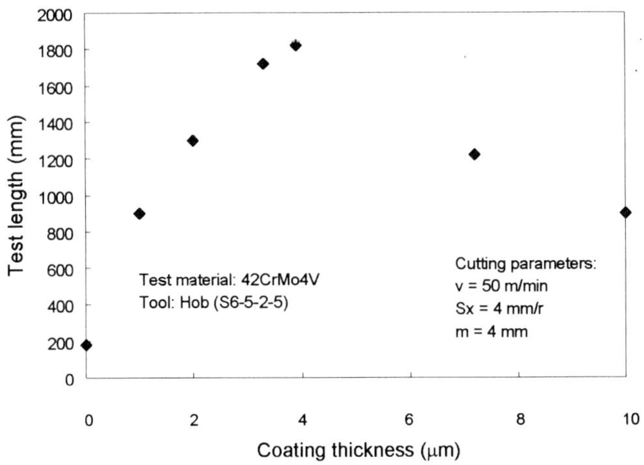

Figure 2.20 Lifetime of TiN-coated HSS hobs as a function of coating thickness.

Other applications

Schiller et al. [13] refer to applications in nine branches of industry, seven of which are given in Table 2.3. Not shown are those in the microelectronics and hybrid microelectronics industries, which seem to have been more progressive than other industrial sectors in exploiting the benefits of PVD. The table gives an indication of the vast potential of PVD in some sectors other than cutting tools, although even this list is not complete.

Work has been carried out into the use of electron beam PVD for the production of thermal barrier coatings for aero engines and other high temperature applications [37,38]. Currently plasma spraying is the method used to deposit these films, but that has inferior adhesion compared to ionisation-assisted PVD techniques.

Electron beam evaporation of aluminium has even been used for the large scale coating of strip steel on a continuous basis [13]. This can be expected to replace conventional tin plating. It can deposit multiple layers, produces no effluents and

Table 2.3 Examples of PVD coatings and applications in various industries

Industrial sector	Layer function	Film material	Product examples
Engineering	Surface protection Wear protection Friction reduction	Al, CrNi TiN, TiC, TaC, WC, Al_2O_3 Oxynitrides, carbonitrides Ag, Mo_2S, C	Machine parts, tool drills, bearing dies
Automotive engineering	Heat/light reflection Friction reduction	TiN, Cr, Al	Automotive
Electronics	Adhesion layer Resistance Conductors and contacts Protective layer Reflection Magnetic store	NiCr, Cr CrNi, Ta_2N, TaAl, Pt, CrSiW, CrSiAl, Cu, FeNi, CuNi, Al, SiO_2, Ta_2O_5, Al_2O_3, $CrSiO_x$, Al Co, Cr, CoCr, Ni	Capacitors, chip resistors, PTC resistors, thin-film transducers, strip-lines. Audio & video discs, storage discs and tapes
Solar	Adsorption, function layer	Me/Oxide–Me–Me/Oxide Cd_2SnO_4, ITO	Solar absorbers, solar cells
Glass	Reflection Selective IR reflection	Al, Al–Ti, CuSn, Ti, CuZn, Me/Oxide–Me–Me/Oxide	Mirrors, architectural glass
Optical	Reflection selection	SiO_2, TiO_2	Optical glasses
Packaging	Decoration	Al, Cr, Cr-alloys, Cu-alloys, TiN, gold silver	Plastics and metals parts. Foils, paper

is very flexible. In the last regard, the electron beams can be used also to heat and degas the steel, and the system can be run as a flow line.

Another feature of the ionisation-assisted PVD techniques which has not yet been widely realised is that they offer the potential to produce duplex treatments and coatings in one process cycle. Thus, for example, plasma nitriding can be carried out, followed immediately by titanium nitride coating [39]. Also the process can be used to produce layers of very different materials. For example, titanium nitride (which has a golden colour) can be deposited and followed by a 'flash' of gold, to seal any pores and allow the product to be sold as gold plated. Most of the commercial systems for producing this type of coating utilize '*load lock*' chambers, whereby the actual deposition vessel can be kept under vacuum at all times with consequential improvements in cleanliness. Also this arrangement permits more efficient processing – allowing (for example) the pre-heating stage to be carried out in one chamber whilst coating is carried out in the next. For maximum effect a further chamber can be incorporated to facilitate cooling. This system of progressive work chambers becomes even more efficient when sequential deposition of several materials is to be carried out in different chambers. An interesting departure from this theme is to deposit an initial layer by a different technique, such as a Solution state process, and then to deposit a subsequent layer or layers by PVD. An excellent example is to put titanium nitride on electroless nickel. In that case the pre-heating PVD stage also serves to harden the electroless nickel [40]. This duplex coating can be excellent in applications in which combined corrosion and wear might occur.

The gaseous or vapour state processes, especially PVD, can be seen to offer considerable potential for many reasons. Their penetration into the engineering coatings market is expected to increase considerably, as process developments occur which facilitate continuous and more economical processing [41], and the use of lower cost substrates [42,43].

References

1. James D H, *Surfacing Journal*, **9**, 3, 1978.
2. HMSO, Wear Resistant Surfaces in Engineering, London, 1986.
3. Dawes C, *Surface Engineering*, **7**, 29, 1991.
4. Rickerby D S and Matthews A (Eds), Advanced Surface Coatings: A Handbook of Surface Engineering, Blackie, Glasgow, 1991.
5. Holmberg K and Matthews A, Coatings Tribology: Properties Techniques and Applications in Surface Engineering, Elsevier Sequoia, Amsterdam, 1994.
6. Bedford G M and Richards P J, Proc. Eng. Surf. Conf. London, May 1986, Inst. of Metals, London, 5/1, 1986.
7. Doyle E D and Jewsbury P, Materials Australia, 8, June 1986.
8. Matthews A, in Surface Engineering Practice, Stafford K N, Datta P K and

Gray J S (Eds), Ellis Horwood, Surface Engineering Practice p33–44, 1990.

9. Lux B and Haubner R, Diamond as a Wear-Resistant Coating, in Diamond Films and Coatings, R. Davis and J. Glass (Eds), Noyes Publications, New York, 1993.
10. Matthews A and Eskilsden S S, Diamond and Related Materials, **3**, p902, 1994.
11. Rossi F, in Advanced Techniques for Surface Engineering, Gissler W and Jehn H A (Eds), Kluwer Academic Publishers, Dordrecht, 1992.
12. Pulker H K, Coatings on Glass, Elsevier, Amsterdam, 1984.
13. Schiller S, Heisig U, Neumann M and Beister G, *Vakuum Technik*, **35**, 35,1986.
14. Moll E and Daxinger H, US Patent 4197, 175, 1980.
15. Morley J R and Smith H R, *J. Vac. Sci. Technol.*, **9**, 1377, 1972.
16. Wan C T, Chambers D L, and Carmichael D C, J. Vac. Sci. Technol., **8**, 99, 1971.
17. Lindfors P A, Mularie W M and Wehner G K, *Surf. Coat. Technol.*, **29**, 275, 1986.
18. Boxman R F and Goldsmith S, *Surf. Coat. Technol.*, **33**, 153, 1987.
19. Munz W D, Hofman D and Hartig K, *Thin Solid Films*, **96**, 79, 1982.
20. Palicki D and Matthews A, *Finishing*, **17**, 36, 1993.
21. Matthews A, Fancey K S, James A S and Leyland A, *Surf. Coat. Technol.*, **61**, 121, 1993.
22. Mattox D M, *Electrochem. Technol.*, **2**, 95, 1964.
23. Berghaus B, UK Patent 510993, 1938.
24. Matthews A, Proc., A Cutting Edge for the 1990's Conference, Sheffield, Sept. 1989, Inst. of Metals, London, 1989.
25. Sjorstrand M E and Thelin A G, Proceedings of the Eighth International Conference on Vacuum Metallurgy, Linz, Austria, Sept. 1985, Eisenhutte Osterreich, 1985.
26. Hatschek R, *American Machinist*, March, 1983.
27. Sproul W D and Rothstein R, *Thin Solid Films*, **126**, 257, 1985.
28. El–Bialy B H, Redford A H and Mills B, *Surface Engineering*, **2**, 29, 1986.
29. Thornley R and Upton D, Proceedings of the Anglo-Finnish Joint Symposium on Advanced Manufacturing Technology, Espoo, Finland, March 1986, VTT, Helsinki, 1986.
30. Clark W, Private Communication.
31. Matthews A and Murawa V, Chartered Mechanical Engineer, **32**, 31, 1985.
32. Tsukamoto T, Sasatis K, Shibuki K, Momma H and Sokichi T, Proceedings of the Conference in Advances in Handmetal Production, Luzern, Switzerland, 1983.
33. Kankaanpaa H and Korhonen A S, *Int. J. Mach. Tools Manufact.*, **27**, 305, 1987.
34. Randhawa H, *J. Vac. Sci. Technol.*, **A4**, 2755, 1986.

35. Posti E, *Tribologia*, **5**, 92, 1986.
36. Konig W, Proc. Triennial Aachen Machine Tool Colloquium, June, p.66, 1986.
37. Fancey K S and Matthews A, *J. Vac. Sci. Technol.*, **A4**, 2656, 1986.
38. James A S, Fancey K S, and Matthews A, *Surface and Coatings Technology*, **32**, 377, 1987.
39. Leyland A, Fancey K S and Matthews A, *Surface Engineering*, **17**, 207, 1991.
40. Leyland A, Bin-Sudin M, James A S, Kalantary M R, Wells P B, Matthews A, Housden J and Garside B, *Surf. Coat. Technol.*, **60**, 474, 1993.
41. Stevenson P and Matthews A, *Surf. Coat. Technol.*, **74/75**, 770, 1995.
42. Matthews A, Artley R J, Holiday P and Stevenson P, The UK Engineering Coatings Industry in 2005, Hull University, 1992.
43. Matthews A and Leyland A, *Surf. Coat. Technol.*, **77**, 88, 1995.

Chapter 3

Electroless nickel coatings: case study

P Gillespie – National Physical Laboratory, UK

3.1 Introduction

The process of electroless deposition is one of the most elegant methods available for the production of alloy coatings. The technique involves the autocatalytic reduction, at the substrate/solution interface, of cations by electrons released from suitable chemical reducing agents and, a significant degree of elemental release from these agents allowing co-deposition with the reduced metal, in order to produce binary, tertiary or even quaternary alloys.

The success of the method stems from the control that can be exercised over the relative strength of the reducing agent in the presence of ligand stabilized metal complexes, allowing alloys of differing composition to be produced. The electroless deposition technique also allows coatings to be produced in amorphous or in microcrystalline non-equilibrium states, each phase exhibiting its own distinct chemical, physical and mechanical properties, whilst heat treatment, subsequent to deposition, can be successfully employed to generate the more traditionally understood equilibrium phases, thereby extending the range of properties available and therefore the number of applications to which the technique can be of use.

The present Chapter undertakes to review the nickel–phosphorus (Ni–P) system, the vast majority of electroless deposition relevant to the industrial sector, world-wide, being largely concerned with the application of this binary alloy to coating technology, although reference will be made, at the conclusion of the article, to other systems of special practical importance.

An account of the historical development of the technique from the early days of its discovery is outlined, followed by a brief description of the essential chemical principals involved in Ni–P deposition, in order both to illustrate the possibilities as well as to outline the flexibility of the process. Consideration is then given to the range of experimental conditions available, as well as to the consequences of the particular choice of these parameters on the resultant alloy composition and therefore on the properties of the eventual coating. The structure of Ni–P alloys is described in terms of the most recently advanced model and this then related to the principal surface properties important to the contemporary coating industry.

Finally, a number of case studies are cited in order to illustrate the reasoning behind the selection of coatings of a particular composition, exhibiting specific properties, required for given working applications.

3.2 The history of electroless deposition

The electroless deposition of metallic nickel from aqueous solution in the presence of hypophosphite was first noted as a chemical curiosity by Wurtz [1,2] as long ago as 1844. Although the metal was almost inevitably precipitated in powder form Breteau [3], in 1911, reported deposits of bright coatings on the surfaces of reaction vessels, which led to the granting of a patent for the chemical process to Roux [4] in 1916.

However, none of this early work led to any practical application and initial development of an industrial process based upon this reaction was not carried out until its fortuitous re-discovery, in 1946, by Brenner and Riddell [5–9]. These workers attributed internal stress of high temperature electrodeposited nickel coatings on nickel–tungsten alloy substrates to the presence of oxidation products derived from the citrate based bath, and in an attempt to relieve this by addition of hypophosphite observed a higher than maximum current efficiency, a phenomenon which they attributed to the action of chemical reduction of the nickel ions. This prompted a full scale development programme by *The General American Transportation Corporation* (GATC) into the possibilities of using electroless plating by chemical reduction alone, as an alternative to conventional electroplating, leading to publication by Gutzeit [10–13], the research director at *GATC*, of a series of papers, between 1959 and 1960, which now form the scientific basis for modern electroless nickel technology.

The early baths developed by Brenner and Riddell were exclusively alkaline in composition, this presumably being a legacy from traditional electrodepositon technology, and necessarily therefore involved the use of large quantities of ammonia. Also the complex reaction chemistry of the deposition process (Sections 3.4 and 3.5) means that electroless deposits are not normally of the pure metal but may also contain small quantities of alloying elements, derived from the reducing agent, such as phosphorus or boron, or elements such as thallium, lead or cadmium derived from other bath additives (Section 3.6).

Early experimental design of electroless baths was devoted to keeping the concentration of these secondary elements to a minimum, which is considerably easier to achieve under alkaline conditions. However, with increasing technological development, from the mid-1950s, the advantages afforded by the use of acidic baths has received greater recognition. These advantages not only include the avoidance of the unpleasant effects of working with large quantities of ammoniacal solution, but it was found that acidic baths tend to be chemically more stable, leading to finer bath control and, generally, to the attainment of higher coating deposition rates.

Further technological advances, in the early 1980's, led to the development of phosphorus rich deposits from acidic baths, as reported by Datta et al. [14–16], giving surface coatings exhibiting far higher wear and corrosion resistance than those produced under alkaline conditions, without the use also of heavy metal bath stabilizers. These coatings are generally reported to form glassy, X-ray amorphous structures [17,18], although more recent work has shown that many have structures consisting of extremely small crystallites, of the order of 50 Ångstroms in diameter, whereas others are believed to be truly amorphous, containing no discernible short range order [19].

3.3 Electroless Ni–P bath chemistry

The chemical deposition of nickel ions on catalytically active surfaces by hypophosphite in aqueous solution can be conveniently described [20] by the following three reactions:

$$Ni^{2+} + H_2PO_2^- + H_2O \rightarrow Ni^0 + H_2PO_3^- + 2H^+ \quad [3.1]$$

$$H_2PO_2^- + H^\cdot \rightarrow P^0 + OH^- + H_2O \quad [3.2]$$

$$H_2PO_2^- + H_2O \rightarrow H_2PO_3^- + H_2 \quad [3.3]$$

These equations describe the net result of a series of simultaneous partial reactions, which have been the subject of extensive mechanistic study [21], yet are still the topic of much discussion. The essential feature of greatest practical importance is that reaction [3.1], which describes the nickel deposition process, takes place at a considerably faster rate than reaction [3.2], which is largely responsible for the deposition of phosphorus, to give nickel rich surface deposits, together with the formation of orthophosphite ion ($H_2PO_3^-$), the overall effect being for bath acidity to increase as the three reactions proceed. Variation of the initial bath conditions permits the relative rates of reactions [3.1] and [3.2] to be controlled thereby allowing, between certain limits, coatings of different chemical composition, and thereby in principle, coatings with different physical and mechanical properties to be produced. Thus, for example, baths designed to work at high pH ranges favour the production of low phosphorus containing coatings, as described in Section 3.6, as well as tending to lead to higher overall rates of deposition.

Similarly, the relative rates of the two reactions, [3.1] and [3.2], may be controlled either by variation in temperature, in the relative initial concentrations of reducing agent to nickel ion, in the choice of both the nature and the composition of added organic complexing agents, as well as in the ratio of substrate area to solution volume within the deposition bath. A full account of all the general factors affecting deposition rate as well as coating composition is given in a comprehensive monograph by Mallory [22], whilst more specific kinetic information together with a review of the mechanisms believed to be responsible for the co-deposition of nickel and phosphorus as well as an account of some practical considerations implicit in the choice of electroless bath conditions, are given in the following three sections.

3.4 The kinetics of Ni–P deposition

A number of studies of the dependency of the rate of deposition of Ni–P coatings have been carried out as a function of temperature and pH, with the influence of effects such as the concurrent production of hypophosphite by-products [23] as well as the estification of complexing ligands also being considered. Mallory, in an extensive study [24], has been able to show that under specific, yet frequently encountered conditions, empirical rate laws describing the kinetics of deposition may be developed. Thus in situations where the solution is sufficiently buffered in order to prevent the acid catalysed estification of the complexing ligands, and where the ligand concentration is not in large (i.e. more than two fold) excess compared to the metal ion concentration, and where the build up of hyperphosphate ion ($H_3PO_3^-$) is small, a situation which is, to all practical purposes true, unless successive bath replenishment is undertaken, then a general empirical rate law may be derived of the following form:

$$\text{Rate} = \frac{d\left[Ni^0\right]}{dt} = k\left[H_2PO_2^-\right]^a\left[N_i^{2+}\right]^b\left[H^+\right]^c e^{\frac{E(T-360)}{T}} \qquad [3.4]$$

The coefficients *a*, *b* and *c* represent the order of the deposition rate with respect to the hypophosphite ion, the nickel ion and the acid concentrations respectively, whilst *E* is the activation energy for the deposition process, *k* the rate constant, and *T* the absolute temperature.

The kinetic description is further simplified by the experimental observation that the reaction orders *a*, *b* and *c* are constant for chelating systems of the same co-ordination number as well as the same annular size. Thus bi-dentate five membered ring complexes formed from glycollate, lactate and aminoacetic anion ligands all follow experimentally observed reaction kinetic behaviour described by equation [3.4], where in all cases $a = 0.8$, $b = 0.6$ and $c = -0.4$. There is however some small observable variation in the activation energies of deposition reactions in the presence of these differing ligands which are found by experiment to fall within the range 17–23 k cal g $mol.^{-1}$, whilst the experimentally determined rate constants, k, vary considerably as a direct function of the particular stability constants between the respective metal/metal-ligand complex of the individual system in question.

3.5 The mechanism of Ni deposition

Although much work has been carried out over the years on mechanistic studies of the Ni–P deposition process, it is widely recognized [20] that the principal reactions responsible for the production of Ni–P surfaces can most accurately be described by either of two mechanistic routes, first proposed a number of years ago.

A: The hydride mechanism

The first step of this proposed mechanism [23,25] is the production of *hydride ion* by catalytic dehydrogenation, at the substrate surface, of the hypophosphite ion, as illustrated in equation [5].

$$H_2PO_2^- + H_2O \rightarrow HPO_3^{2-} + 2H^+ + H^- \quad [3.5]$$

It is suggested that this anion then acts as the principal agent responsible for the reduction of nickel ions in solution, to form a metallic surface deposit, accompanied by the liberation of hydrogen gas, equation [3.6].

$$Ni^{2+} + 2H^- \rightarrow Ni^0 + H_2 \quad [3.6]$$

The presence of a phosphate containing intermediate species, such as metaphosphorus acid (HPO_2), is also proposed, and this it is believed is further reduced by hydride ion at the surface to produce phosphorus metal, again with the evolution of hydrogen gas.

$$2PO_2^- + 6H^- + 4H_2O \rightarrow 2P^0 + 3H_2 + 8OH^- \quad [3.7]$$

B: The Gutzeit mechanism

This mechanism [10–13, 25] suggests that *atomic hydrogen* is the species most active in the promotion of the nickel deposition reaction.

Firstly the hypophosphite ion is catalytically oxidized at the substrate surface.

$$H_2PO_2^- + H_2O \rightarrow H_2PO_3^- + 2H\cdot \quad [3.8]$$

The resulting radicals are thought in turn to reduce the nickel ions to form a surface deposit of the metal, together with the release of acid into solution.

$$Ni^{2+} + 2H\cdot \rightarrow Ni^0 + 2H^+ \quad [3.9]$$

Similarly, some of the hypophosphite ions are, it is proposed, simultaneously reduced by the active hydrogen to deposit phosphorus metal onto the substrate surface.

$$H_2PO_2^- + H\cdot \rightarrow H_2O + OH^- + P^0 \qquad [3.10]$$

It is furthermore proposed, under this mechanism, that surface catalytic oxidation of a proportion of the hypophosphite ion is responsible for the liberation of hydrogen gas.

$$H_2PO_2^- + H_2O \rightarrow H^+ + HPO_3^{2-} + H_2 \qquad [3.11]$$

3.6 Choice of electroless bath conditions: some practical considerations

The historical evolution from the use of alkaline [5–9] to acidic [10–13] electroless deposition baths, Section 3.2, was largely dictated by the need to develop a large scale industrial process and to therefore avoid the disadvantageous health and safety aspects of working with large volumes of ammoniacal solution.

Also it was initially found that the complex reaction chemistry involved in the deposition process meant that electroless deposits are not normally of the pure metal, but consequently include elements derived from the phosphorus and boron containing reducing agents, an essentially fortuitous effect, which has more recently been used to considerable advantage [14–16], enabling selection of certain tribological properties to be performed.

Similarly, it has been established that a number of other important experimental conditions may be varied, within certain constraints, in order to produce coatings exhibiting differing properties. The general requirement of an electroless deposit is that it should provide an integral, smooth and well adhered coating to the chosen substrate, and these conditions in themselves result in the imposition of some constraints in the choice of experimental bath conditions.

Furthermore, other requirements of the properties of the resultant coatings depend very much on their eventual proposed use. The principal concern could, for example, be to produce surfaces exhibiting specific wear or corrosion properties, or be directed towards the development of particular electrical or magnetic characteristics, be aimed at producing coatings of a certain required thickness, or on the other hand, be directed towards the production of a particular aesthetic allure. Thus the development of the electroless deposition process has led to much experimentation, with a view to obtaining a fine balance between the choice and relative composition of the principal chemical components of a bath, together with the relationship between these and physical parameters, such as temperature, pH and bath loading values, experiments often resulting in final optimisation of bath conditions by the inclusion of stabilizing and rate accelerating agents.

In principle, the properties of as-deposited coatings depend very strongly upon their phosphorus content, whilst strong experimental evidence has been obtained [26,27] to suggest that the rate of deposition is inversely proportional to the phosphorus content of the resultant coating, and although it is clear that this effect is more pronounced for certain variable experimental bath parameters than for others, it is considered to be a good working hypothesis to assume that variation in any parameter that results in an increase/decrease in deposition rate will tend, albeit to varying degrees, to lead to a decrease/increase in the phosphorus content of the same coating. The principal experimental bath parameters and their influence on the properties of resulting surfaces are now discussed, with particular attention being paid to the effects of these on both coating composition and relative rates of deposition.

3.6.1 Bath composition

By far the most important single parameter in determining the overall composition of coatings is the ratio between the concentration of nickel ions in solution to that of the phosphorus containing reducing agent. It is generally agreed, as first proposed by Gutzeit and Krieg [28], that this ratio should lie between 0.30 and 0.45 for nickel ions in the presence of a hypophosphite reducing agent, the higher this value the lower the phosphorus content of the resulting coating. Compositional ratios below 0.25 have been shown to lead to the production of coatings containing undesirable brown deposits due to the presence of excess phosphorus, whilst ratios larger than 0.45 lead to very slow reduction, and consequently to the production of thin coatings due to the lack of sufficient reducing agent [$H_2PO_2^-$].

In principle, any soluble ionic nickel salt could be chosen as a source of metal ions, although economic considerations have meant that by far the majority of work has been carried out with either the chloride or the sulphate, the sulphate being preferred in recent years, particularly when working under acidic conditions, in order to avoid the potential detrimental effects caused by the long term build up of chloride ion in the deposition bath. Similarly, for economic reasons as well as for reasons of availability, the choice for the source of hypophosphite ion is inevitably the sodium salt.

3.6.2 Complexing agents

Whilst complexing agents exert a considerable buffering action, enabling bath acidity to be controlled during the deposition process, the principal rôle of these agents is to combine chemically with otherwise free nickel ions in solution, thus helping to stabilize the bath and aiding in the prevention of spontaneous reduction. It is of paramount importance to chose both an agent and experimental conditions that will generate a complexing ligand that is neither too strong in its chelating ability, thereby stabilizing the nickel ion to such an extent as to render the deposition rate impractically slow, nor so weak that its chelating ability is ineffectual, resulting in its performance being masked by the action of other buffering ions in the system.

In the kinetic description of the electroless deposition process described by Mallory [24] Section 3.3, conditions were chosen in order that the rate of deposition would prove to be independent of ligand concentration, this corresponding to the ideal condition of intermediate ligand strength, which has been found by experiment to correspond to situations where the ligand concentration at all times is greater than one third, but less than one and a half times, the nickel ion concentration. Work has been carried out with a large number of ligands, the most popular being dicarboxylic acids, which act, under intermediate conditions, as bi-dentate five membered ring chelates. Many other systems [29] have been successfully employed, however, particularly those based upon mono carboxylic acids as well as both α and β aminocarboxylic acids, whilst much attention has also been paid to the use of mixtures of a variety of strong and weak ligands in order to set up equilibrium conditions enabling the joint advantages of the presence of both types of ligand to be attained, thus minimising the amount of free nickel ion in solution, yet maintaining an experimentally viable deposition rate.

3.6.3 Stabilizers

The primary function of stabilizers is to prevent alloy deposition on the walls of the reaction vessel by effectively poisoning the catalytic activity at these sites, whilst their presence has often been reported to have a small beneficial effect on the rate of deposition as well as giving coatings that are much brighter in appearance.

The most commonly used stabilizers are lead, cadmium and thallium containing salts or alternatively thiourea, all of which are added in concentrations of the order of 1–2 mg. l^{-1}, in order to not extend the poisoning effect to the whole of the deposition bath. Organic stabilizers in general, such as thiourea, thioglycollic acid and cystine have a tendency to lower the phosphorus content of coatings due to their presumed interaction with the hypophosphite ion, whilst some recent concern has been expressed about the use of heavy metal ions, especially for applications relating to the food industry, as these ions are often found to be co-deposited along with nickel in amounts disproportionate to their concentrations in solution.

3.6.4 Acidity

The effect of pH on the deposition rate of nickel can be appreciated by considering the effect of the factor $[H^+]^c$ in expression [3.4], derived by Mallory [24], Section 3.4, for the case of intermediate complexor strength and where the nickel to hypophosphite concentration ratio lies between 0.30–0.45. Under these recommended conditions the dependency of the kinetics of the system on acid concentration can be successfully described by a constant value for b of – 0.4 , showing that an increase in acid concentration (decrease in pH) has a retarding effect on the rate of deposition.

Also it has been shown by experiment that for pH values lower than 4.5 the deposition reaction is too slow in order to obtain large deposits over a reasonable time scale whilst pH values greater than 5.2 lead to very fast deposition, which has been shown, on the other hand, to be responsible for the production of Ni_2P inclusions, which manifest themselves in the form of undesirable surface pinholes [29].

3.6.5 Temperature

The temperature dependency of the rate of deposition described by the term [exp E (360–T)/T], Section 3.4, indicates a very strong increase in the nickel deposition rate with rising temperature to such an extent that a temperature rise of between 75–85°C, will have the effect of doubling the reaction rate. In practice it is generally agreed that, unless there is some special reason not to, such as the nature of the substrate, deposition should be carried out at the highest temperature that is practically possible, and that chemical control, such as the use of complexing agents, stabilizers and buffers, should then be specifically employed in order to reduce the risks of instability associated with high temperature baths.

3.6.6 Bath loading

The bath loading parameter describes the ratio between the surface area of work exposed to treatment compared to the total volume of electroless solution contained in the plating bath. It is of particular importance, in the case of first and second generation Ni–P electroless deposition bath systems, as it has a major influence on the rate of deposition, as well as therefore, on the phosphorus content of the resultant coating. Table 3.1 illustrates the comparative rate of growth (measured

Table 3.1 Deposit thickness as (in μm) a function of time for different bath loading parameters

	Bath loading ($cm^2 l^{-1}$)			
Time (hours)	**50**	**100**	**200**	**400**
1	25	19.5	16.2	8.8
2	43.5	21.0	25.1	13.8
3	57.0	45.0	30.2	16.5
4	67.5	52.5	33.8	17.9
5	77.4	58.1	34.9	19.3
6	82.5	60.7	37.2	20.0

as deposit thickness, in microns) of Ni–P electroless deposits for a typical first generation bath (one that has not undergone replenishment) for different loadings at constant temperature.

In practical terms it can be seen that coating thickness in all cases tends towards a limit, there being an inverse relationship between this limiting value for any given situation and the corresponding value of the bath loading parameter. This gradual reduction in rate of deposition with time is primarily due to the decrease in concentration of reactants in solution as the coating is formed and conforms to the empirical kinetic law governing rate of deposition as described in Section 3.4 (equation [3.4]). Clearly some consideration should be given therefore to the most appropriate bath loading, value based upon the eventual thickness of coating required, it being impractical, for example, to attempt to coat to a thickness of 40 mm under the experimental bath conditions employed in Table 3.3, with a bath loading parameter of 200 cm^2 1^{-1} or greater. Any restrictions that the relationship between bath loading parameter and rate may appear, at first instance, to impose upon the phosphorus content of the coating may be overcome by modification of the bath composition, remembering to maintain the ratio of nickel ions to hypophosphite ions to between 0.30–0.45, in order to avoid problems of either slow reduction or discoloured deposits, or to modify the actual concentrations of the reactants,or alternatively, to change entirely the nature of the complexing agent system.

3.7 Structure of alloys

Traditionally, composition or phase diagrams are based upon the results of experiments performed on alloys formed from slow solidification from the melt, prepared in situ from the pure component elements, to form phases that are in thermodynamic equilibrium. A phase diagram for the binary nickel-phosphorus system was first reported by Konstantinov [30] in 1908, many years before the development of electrolessly produced nickel, and was subsequently modified by Hansen in 1943 [31], and then again by Metcalfe in 1958 [32].

The principal features of the nickel-rich corner of the binary Ni–P phase diagram, Fig. 3.1, include the first eutectic point, (880°C) at 11.0 wt. % P, as well as compound formation at compositions corresponding to Ni_3P (1025°C), Ni_5P_2 (1175°C) and Ni_2P (1110°C). A small solid solution region, designated α-nickel phase, has been shown to contain 0.17 wt. % P, and this, together with the intermetallic Ni_3P compound, are the only two phases reported in this region.

An extensive amount of research has been carried out, over the last fifteen years, on the production of Ni–P coatings by electroless deposition, and has indicated that the properties of the resultant surfaces often cannot be described by this phase diagram alone, and although the overall picture has for a long time remained confused a great deal of recent progress has been made in the

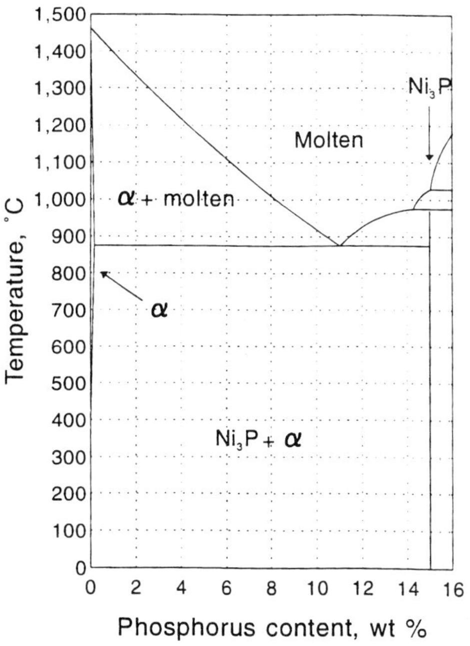

Figure 3.1 Phase diagram for nickel-phosphorus alloys.

understanding of many aspects of the system.

For example, one of the more detailed studies reported to date [33] describes the electronic structure of amorphous Ni–P metallic glasses of composition $Ni_{89}P_{11}$ and $Ni_{79}P_{21}$, prepared by rapid quenching and studied by X-ray photoelectron spectroscopy. Results of this study indicate that the structure varies from a microcrystalline deposit ($Ni_{89}P_{11}$) with very short range order, where the phosphorus atoms are surrounded by nine nearest nickel atom neighbours, in a supersaturated solution of phosphorus within a distorted face centred cubic nickel lattice, to a composition ($Ni_{79}P_{21}$), very close to the eutectic point, Fig.3.1, where the local environment of the phosphorus atoms contains no short range order, the alloy thereby being said to be truly in the "glassy" amorphous state. Furthermore, the authors report that annealing of these two non-equilibrium phases leads to the identification of phase transition temperatures in the region of 350°C, above which the results of further spectroscopic studies suggest that Ni_3P and Ni crystallisation is favoured.

Results of other studies [34] confirm that as-deposited Ni–P coatings can have structures ranging from extremely small crystallites of the order of 5 nm to those which are totally amorphous, the most recent results [35] for deposits containing between 3.0–7.3 wt. % P indicating that the particle size lies within the range 1.4 to 11.9 nm. Furthermore, the problems associated with X-ray measurements to determine crystallite size have been examined in detail by Kreye et al. [36], demonstrating at the same time the power of the XRD technique in observing amorphous-to-crystalline transformations, an amorphous 12.0 wt. % P alloy being clearly shown to transform into Ni_3P and α-nickel on exposure to 400°C for a period of one hour.

In an extension of this work [37] electron diffraction (ED), transmission electron microscopy (TEM) and X-ray diffraction (XRD) results for a series of heat treated as-deposited Ni–P alloys within the compositional range 2.3–14.0 wt. % P. were reported. The grain boundary of low phosphorus, (i.e. between 2.3–4.0 wt. % P in this case) microcrystalline compositions was estimated by TEM to be of the order of 2 nm, whilst high phosphorus containing alloys, > 11.0 wt. % P, were shown by both ED and XRD to be totally amorphous in nature. Further XRD results of annealed samples indicated that the formation of crystalline nickel occurs at temperatures inferior to that at which formation of Ni_3P begins, for all compositions up to 11.0 wt. % P, after which the two reactions were observed to proceed simultaneously.

Many other structural studies of the metastable non-equilibrium states have confirmed that these largely consist of either finely divided or amorphous material and that heat treatment of these often results in the identification of glass transition temperatures, followed by the appearence of equilibrium phases as described by the traditional phase diagram.

It has also been suggested by Riedel [25], in a review of studies carried out by a number of workers, that in order to clarify the situation, it should be considered that two temperature ranges are of importance so far as thermally induced modifications of the physical and tribological properties of electrolessly deposited surfaces is concerned, and that, as well as the frequently reported high temperature structural modifications, a "baking" treatment, at relatively modest temperatures up to 280°C, may under certain situations change the properties of alloys without any structural changes being observed, this being particularly true for the hardening of low phosphorus deposits (» 2.3 wt. % P) as reported by Parker [39].

An extremely important recent advance in the understanding of many of the results concerning studies of the properties of electrolessly deposited coatings has however been gained from the publication by Duncan [40] of a compositional diagram, Fig. 3.2, designed to describe what often seems to be the complex relationships between metastable and equilibrium states, such as those described above. This diagram represents a comparison of results obtained from studies of electroless coatings investigated by a number of different workers, using a variety of techniques [41–52], with those of the traditional phase diagram [32], Fig. 3.1, which describes the results of equilibrium phases obtained by cooling of compositions from the binary melt. Duncan's compositional diagram not only predicts the nature of the as-deposited structure of alloys formed from the electroless technique in terms of two new phases, designated β and γ, but also indicates the temperature as well as the compositional range within which β-to-γ

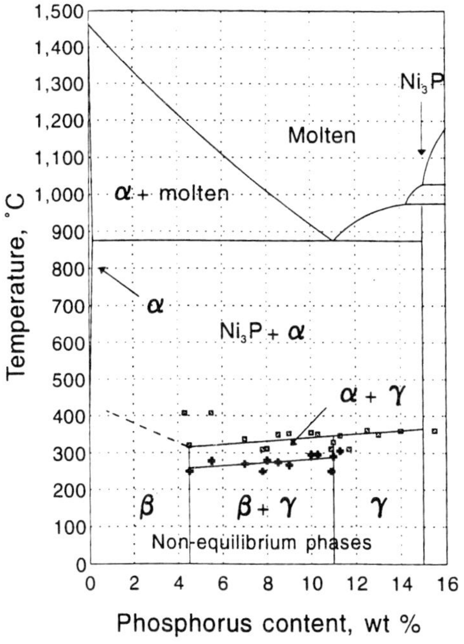

Figure 3.2 Phase diagram for electroless nickel deposits.

and γ-to-Ni_3P phase transitions may be expected to occur.

Thus low phosphorus as-deposited coatings consist solely of the non-stoichiometric β phase, which is described as a crystalline solid solution of phosphorus in nickel, similar to the α-nickel phase reported in the equilibrium phase diagram, Fig. 3.2, but containing up to 4.3 wt % P. High phosphorus coatings, i.e. > 11.0 wt % P, on the other hand, form a totally amorphous γ phase, which may contain up to 15.0 wt% P, whilst compositions containing between 4.3–11.0 wt. % P consist of mixtures, in various proportions, of β and γ phases.

Controlled heat treatment of the β phase will induce the formation of the α-nickel phase, together with crystalline Ni_3P at 310°C, although much higher temperatures may be required to initiate this reaction for alloys of very low phosphorus composition. Similarly, the amorphous γ phase will decompose when heated above 330°C for 11.0 wt % P, but not until above 360°C for 15.0 wt % P, again with the formation of α-nickel and Ni_3P equilibrium phases.

Alloys of intermediate composition, comprising β and γ as-deposited meta-stable phases, undergo a β-to-α transition when heated to above 250–290°C, depending on the exact alloy composition, to form α and γ phases, whilst further heat treatment to above 310–330°C results in conversion to the α-nickel and crystalline Ni_3P equilibrium phases.

To summarize, the volume of evidence supporting the existence of metastable phases is overwhelming, their importance being enhanced in the particular case of electroless coatings, as it is these phases that are more often preferentially formed by this method of deposition, as opposed to the equilibrium phases normally associated with structures obtained from traditional cooling methods, whilst the subsequent thermal manipulation of these non-equilibrium phases contributes to a significant extent to the interesting and specific chemical, physical and tribological properties of electrolessly formed Ni–P coatings.

3.8 Physical properties of Ni–P coatings

Recent advances in the understanding of the nature and structure of as-deposited as well as heat treated Ni–P alloys, as described in the previous section, will have, it is predicted, far-reaching consequences on the future development of electrolessly produced coatings. The primary concern, in practice, is to choose experimental bath conditions with a view essentially to maintain high as well as uniform plating rates, thus creating experimental conditions capable of depositing high quality Ni–P coatings on a complete range of articles and artefacts.

Within the constraints of these primary conditions it has been shown that not only can binary alloys of different composition be produced, but some considerable degree of flexibility in chosen experimental bath conditions can be exercised in order to permit alloy composition to be controlled to a large degree of precision.

Relationships between various tribological properties of the electrolessly deposited Ni–P alloys and their nominal composition have long since been recognized and have led to general acceptance of the classification of these alloys into four principal groups, described as low phosphorus (1.0–3.0 wt % P), low–medium phosphorus (3.0–6.0 wt % P), medium phosphorus (6.0 –9.0 wt %) and high phosphorus (9.0–13.0 wt % P) alloys.

The often generally distinct physical properties of alloys belonging to these designated groups can be understood in terms of the various structures formed in each individual case, as described in Section 3.7, and can be seen to be paramount to the choice of a particular alloy composition for any specific given application.

3.8.1 Hardness

Traditionally, one of the most important and frequently quoted tribological properties of Ni–P surface coatings is the Knopp microhardness. The compositional and temperature dependent values of Knopp hardness values of Ni–P alloys are compared in Table 3.2 with recently proposed structural descriptions, Fig. 3.2, where it can be seen that the occurrence of dramatic changes in hardness tends to coincide to a large extent to transitions within the alloy structures.

Thus the microhardness of coatings that have not been heated beyond 250°C can be seen to be significantly reduced from 728–838 for low phosphorus (1.0–

3.0 wt % P), which consist purely of the β phase, to values ranging from 513 to 643 for medium phosphorus containing alloys (3.0–9.0 wt % P), consisting of mixtures of β and α phases, and then further reduced to values ranging from 517 to 572 for high phosphorus coatings which consist purely of the amorphous γ phase. Similarly the marked increase in experimentally determined microhardness of low phosphorus containing alloys (1.0–3.0 wt % P) from 728–838, to give hardness values of between 927–987 on heating to between 350–400°C can be seen to be coincident with an expected β → (Ni_3P + α) transition, whilst increases in the microhardness of both medium (3.0–9.0 wt. % P) and high (9.0–13.0 % P) phosphorus containing alloys on heating can be related to (β + α) → (Ni_3P + α) and to γ → (Ni_3P + α) phase transitions respectively.

3.8.2 Wear

The wear resistance of electroless Ni–P alloys is greatly influenced by the particular operating mechanisms, and although surface hardness has often been found to be an extremely useful indicator to wear performance this parameter should not, in the case of Ni–P coatings, be used uniquely, to determine the resilience of a coating under specific conditions.

One of the most frequent situations encountered in practice involves the evaluation of abrasive wear, which can be defined as the removal of material by either cutting or ploughing from the surface by a harder material under load. Thus the *Taber wear index* (TWI), where the surface weight loss is reported under standard conditions, is the most commonly employed indicator of the relative

Table 3.2 The Knopp hardness values of Ni-P deposits of different composition, before and after heat treatment, compared to identified phases

	Phosphorus content		
	Low (1.0–3.0 %)	**Medium (3.0–9.0 %)**	**High (9.0–13.0 % P)**
As-plated	728–802 β	513–527 β+γ	517–531 γ
250°C	728–838 β	612–643 β+γ	560–572 γ
300°C	795–877 β	872–918 α+γ	870–918 γ
350°C	980–985 Ni_3P + α	862–952 Ni_3P + α	856–977 γ
400°C	927–987 Ni_3P + α	834–897 Ni_3P + α	866–987 Ni_3P + α
450°C	854–860 Ni_3P + α	800–934 Ni_3P + α	811–894 Ni_3P + α

resilience to wear of a given surface. Experimentally determined results indicate that wear measured by this type of empirical experiment does generally exhibit a proportional relationship to Knopp microhardness. In the case of as-deposited Ni–P electroless coatings a near linear relationship between alloy composition and wear resistance is usually reported, low–medium phosphorus containing coatings (3.0–6.0 wt. % P) giving TWI values of between 4–10 mg/1000 cycles under a load of 10N, rising to around 25 mg/1000 cycles for 11 wt. % P coatings, medium phosphorus containing coatings (6.0–9.0 wt. % P), on the other hand, generally exhibiting TWI values of the order of 12–19 mg/1000 cycles under the same conditions of load. Heat treatment has the effect of improving wear resistance for alloys of all composition, annealing at 300–350°C, in order principally to promote the $(\beta + \gamma) \rightarrow (\alpha + Ni_3P)$ phase transition, having the effect of decreasing TWI values by as much as 25%.

Adhesive wear, on the other hand, can be defined as removal of material between interacting surfaces as a result of the adhesion of asperities, and is characterized by the transfer of material from one surface to another. It has been shown that in the case of Ni–P alloys, adhesive wear resistance, contrary to abrasive wear, improves with increasing phosphorus content.

Generally however a wear experiment should be designed with a view to reproducing, as close as possible, the conditions encountered in practice during the working life of the surface, should this be wear against a particular substrate, high temperature wear, or wear in the presence of particular fluids or gases.

3.8.3 Corrosion protection

The level of corrosion protection provided by a particular Ni–P alloy depends upon its ability to form integral low porosity coatings as well as upon the extent of its chemical inertness to the immediate working environment.

Whilst porosity has been found by experience to decrease rapidly with coating thickncss for alloys of all composition, those with phosphorus contents greater than 10.0 wt. % have been found to exhibit particularly low porosity, so more readily fulfil therefore the essential requirement of providing protection by the formation of an integral protective barrier coating. Heat treatment generally is found to be detrimental to corrosion protection due to the formation of microcracks formed through inevitable mismatches in the physical properties of phases present before and after thermal treatment.

The relative degree of chemical protection afforded to alloys of different composition is a direct function of the reactivity of the particular alloy under different chemical environments. Acid attack is generally more prevalent in the case of low phosphorus containing alloys, although in contrast these alloys may provide protection, and find use therefore, in work involving alkaline substances. A notable exception to this observation is the case of ammoniacal based solutions, which readily form soluble complexes with nickel, Ni–P alloys, particularly those

Table 3.3 Magnetic properties of Ni–P alloys of different composition

Composition	Structure	Coersivity (Oe)
3.0–6.0	High β content	80–20
7.0–9.0	($\beta + \gamma$)	1–2
11.0	Pure γ	0

low in phosphorus, being particularly unsuitable therefore for applications involving contact with these types of substances.

Mineral acids such as hydrochloric and nitric acid exhibit corrosion properties that are strongly dependent upon alloy composition, these two reagents in particular being used as standard qualitative tests for evaluation of relative corrosion resistance of Ni–P alloys. Many organic acids, such as acetic, readily form stable complexes with nickel yet are inert in the presence of common substrate materials such as copper, aluminium and steel, so that whilst Ni–P alloys are unsuitable for applications involving contact with solutions of this type the latter, incidentally, have important rôles to play as striping agents for Ni–P coatings in the presence of these substrates [53, 54].

The resilience of high phosphorus containing alloys (> 10.0 wt. % P) to mineral acid and to salt spray tests, as well as the ease of which these alloys form coatings of low porosity, leads generally to these compositions being preferred in applications where good corrosion resistance is the essential requirement. However, as illustrated in the following section, in practice, situations may occur where a degree of corrosion protection may be successfully sacrificed in order to obtain benefits in other physical or mechanical properties attainable only by use of lower phosphorus content alloys.

3.8.4 Stress

Probably the most striking correlation between any given physical or mechanical property of Ni–P coatings and alloy composition can be obtained by measurement of the internal stress of the as-deposited coating and the phosphorus content, as illustrated in Fig. 3.3.

The points which define the transition between compressive and tensile stress are estimated to occur at 3.6 wt. % and at 10.4 wt. % phosphorus, between which

Table 3.4 Specific resistivity of Ni–P alloys of different composition

Composition (wt % P)	Specific resistivity (Ω cm x 10^6)
Pure Ni metal	6
3.0	30
5.0–6.0	72 (reduced to 20 on heating to 600°C)
8.0–9.0	30–100

coatings are deposited under tensile stress. This region corresponds well with the compositions between which as-deposited coatings consist of mixtures, in various proportions, of β and of γ phases (i.e. between 4.3–11.0 wt. % P), Fig. 3.2. The tensile, rather than the compressive nature of the force within this region is found to be detrimental to the fatigue life of the coating, tending to enlarge rather than replenish wear or age related microfractures. Conversely, fatigue life is found by experience to be extended under the influence of compressive stress, in situations where coatings consist of pure phases, whether these be exclusively the pure β phase, or alternatively, the pure γ phase.

3.8.5 Magnetic properties

The equilibrium α-nickel phase is found, like pure nickel metal, to be strongly ferromagnetic, whilst the magnetic coersivity of Ni–P alloys has been shown to decrease with increasing phosphorus content as shown in Table 3.3, high phosphorus containing alloys (> 11.0 wt. % P) exhibiting non-magnetic behaviour [55,56]. Heat treatment of medium phosphorus (6.0–9.0 wt. % P), on the other hand, has been shown to lead to an increase in magnetic coersivity up to between 100–300 Oe, due to the effect of the (β + α) → (α + γ) thermal phase transition.

3.8.6 Electrical properties

The absolute measurement of the specific resistivity of Ni–P alloys may often be complicated by the presence of, as well as the time dependent release of, chemi-

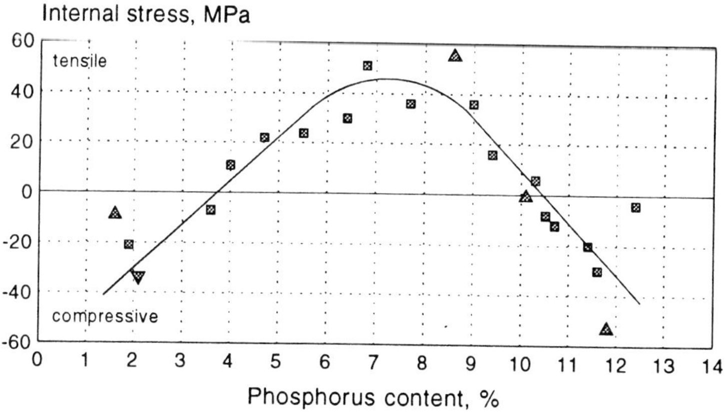

Figure 3.3 Effect of phosphorus content on internal stress.

adsorbed hydrogen, as well as by longer term thermodynamic equilibration of the non-equilibrium as-deposited phases.

The generally agreed ranges for experimentally determined specific resistivity of as-deposited alloys of differing composition are compared in Table 3.4 with the value for pure nickel metal [38].

The three-fold reduction in resistivity found on heating 5.0–6.0 wt. % P alloy can be explained in terms of the $(\beta + \gamma) \rightarrow (\alpha + Ni_3P)$ phase transition, the electronic properties of α-nickel, containing only up to 0.17 wt. % P, being similar to that of the pure metal.

3.9 Application of electroless Ni–P coatings: case studies

It can be seen, as described in the previous section, that considerable flexibility in the properties of electrolessly deposited surfaces can be achieved by a combination of careful control of the experimental bath parameters, followed by application of well defined heat treatment procedures on the deposited coatings. Thus surfaces can be produced with preferentially enhanced properties, chosen in order to suit the specific major requirement of a particular application. This point can be conveniently illustrated by the following descriptive accounts of a number of selected case studies:

Study 1: Slide tracks for mini van doors

The working surface, in this application, is exposed to repetitive wear as well as to typical roadside corrosion environments. The surface coating protecting the tracks is chosen, in principle, so as to prevent the formation of corrosion products which would impair the long term free operation of the system. In order to gain this, some hardness or wear resistance is sacrificed in order to achieve the high corrosion protection required for this particular application. Thus high phosphorus (i.e. > 10.0 % P) coatings are preferred as they generally possess the greater ability to form integral coatings with a hermetic barrier, and are therefore very much lower in porosity, a factor that greatly assists general corrosion resistance. Similarly the extra hardness that would, in principle, be provided by heat treatment is rejected due to the accompanying increase in porosity that would occur if these articles were heat treated.

Study 2: Differential pinion shafts

Very good wear resistance is essential for this application, whereas good corrosion protection is of very much less importance, which operates in a protective environment of oil. In this instance both low and medium phosphorus deposits are suitable followed by post heat treatment to 400°C for at least one hour in order to promote maximum surface hardness.

Study 3: Injection mould die cavities

The principal rôle of the surface coating in this application is to protect the internal cooling tubes from the liquid coolant, which would, if the two were to come together, severely reduce the lifetime of the die. High wear and high acid corrosion performance are not particularly important therefore for this application. Furthermore, the large metal moulds, if not pre-warmed before plating, require considerable time to achieve and maintain a stable plating temperature, so that these articles lend themselves to the application of low phosphorus coatings as these can be plated at lower bath operating temperatures.

Study 4: Oil industry components

In this industry different requirements are often sought, depending upon the particular application, and choice is therefore often made between coatings exhibiting either high corrosion resistance or high hardness. For example, in certain field situations coatings are required in order to protect items from the particularly corrosive conditions normally generated in the hostile environments encountered at sea. Thus thick (» 100 mm) high phosphorus as-deposited coatings are chosen to ensure long service life of equipment used for oil/sea-water separation processes.

But in other situations frequently encountered in the oil refining industry, such as where highly abrasive sand and rock containing slurries need to be pumped through pipes and valves, good wear resistance rather than high corrosion resistance is required, and relatively thin (20.0–50.0 μm) heat treated medium phosphorus (3.0–9.0 wt % P) coatings are considered to be most suitable for this particular application.

Study 5: Rail tankers for caustic liquor transportation

One of the first commercial applications of electroless nickel coatings was for the protection of steel tankers used in the transport by rail of strong caustic liquors. Whilst under conditions of general environmental corrosion high phosphorus containing coatings are found to give maximum protection this is essentially under acidic conditions. It has however been found from experience that the reverse is true for chloro-alkaline conditions, where low phosphorus containing as-deposited coatings give the most efficient protection against corrosion.

Study 6: Thin film memory discs

The electronics industry is responsible world-wide for approximately 35% of Ni–P electroless deposition technology, the majority of this work being applied to the manufacture of thin film memory discs for use in the computer industry.

In this process a thin highly uniform temperature stable coating is required for application onto aluminium disc substrates in order to enable thin layers of cobalt and then graphite to be successively applied by the method of high temperature (250°C) sputtering.

Attention was initially drawn to electroless nickel–phosphorus for use as the intermediate layer in this application due to its ability to form highly uniform, dense coatings compared to other methods of deposition, as little variation in thickness of these high precision components can be tolerated.

However, other technological requirements have to be met before a material can be considered suitable for this application. The initially applied as-deposited electroless coating is normally 12–15 μm thick and needs to undergo fine polishing to produce a surface finish of mirror quality, and in order to achieve this degree of precision maximum hardness of the initial deposit is required, thereby eliminating the use of low phosphorus containing coatings for this application.

Furthermore, the electroless coating, to be of potential use in this application, should be able to withstand sputtering temperatures of 250°C for periods of up to one hour, and should also exhibit non-magnetic behaviour in order to prevent interference of the magnetic switching process between the read/write head and the cobalt data storage film. The choice of high phosphorus (> 11.0 wt % P) containing coatings is therefore ideal for this application as it provides hard, temperature stable, non-magnetic coatings.

3.10 Other specialist applications

Whereas Ni–P electrolessly deposited alloys play a predominant rôle in coating technology other compositions are available, which, although they cater for only a small proportion of the market, do considerably broaden the range of applications to which electroless nickel coatings may serve.

3.10.1 Nickel–boron (Ni–B) coatings

The use of Ni–B, as opposed to Ni–P coatings, as inert protective barriers for steel substrates was first proposed in 1954 [57] and has been the subject of a more recent study [58] involving its more general applications. Chemically, the technique is analogous to that of Ni–P deposition, elemental boron however, rather than phosphorus, being introduced into the deposit from the reducing agent, the extent to which this occurs again being the primary factor responsible for determining the physical properties and performance of the resultant coating. The reductant normally employed in electroless Ni–B technology is usually either sodium hydroboride or, more extensively in recent years, any of a range of the more water soluble organic aminoboranes such as for example dimethylamine borane (DMAB).

For the purpose of definition, based upon their distinctive tribological properties, alloys are normally described either as low boron containing, with between 0.2–0.3 wt. % boron, or as high boron containing, where as much as 5.0 wt. % B may be present in the final alloy.

Low boron deposits generally exhibit very good electrical properties, with low contact resistance and exhibiting relatively high electrical conductivity. This, coupled with the ease with which they can be soldered as well as their good storage-life solderability means that they have found many applications in the electronics industry, in the plating of metallized ceramics, in the development of hermetically sealed packaging and as a cost effective undercoat for gold surfaces.

High boron alloys, on the other hand, are characterized by their good hardness properties, particularly at high temperature and more specifically still when involved in coating applications to materials that would otherwise be in contact with molten glass. Their use, in this context, has been demonstrated to great effect for the protection of low cost aluminium guides employed in the transference of molten metal from furnaces, as well as in the bottle making industry where these types of coatings have been found to greatly increase the lifetime of bottle moulds, which due to the high temperatures involved, are necessarily fabricated in steel.

3.10.2 Electroless Ni/PTFE

The inclusion of PTFE particles into traditional electroless deposits has the effect of extending the coating low friction coefficients, this having a strong advantageous effect in low load, high slide wear applications, where excellent wear resistance is the primary criterion for the choice of coating. This advantage is, unfortunately, offset by a substantial reduction in corrosion protection and often leads therefore to PTFE enhanced electroless nickel being deposited onto more conventional nickel surfaces. Applications of this type of coating include water pump shafts as well as steel and aluminium components for air conditioning units, for example, where low noise and therefore low friction components are required.

3.10.3 Electroless Ni/hard particles

The inclusion of particles having well documented high hardness properties into electroless nickel coatings can be effected in order specifically to produce surfaces with extended wear protection. Thus particles of silicon carbide, tungsten carbide and diamond can be entrapped within electrolessly deposited matrices in order to impart reinforced wear properties required for certain specialist cutting tool applications.

References

1. Wurtz A, *C R hedb. Seances Acad. Sci.* **18**, p702, 1844.
2. Ibid. **21**, p149, 1845.
3 Breteau P, *Bull. Soc. Chim.* **9**, p515–518, 1911.
4. Roux P US Pat. 1 207.218 (1916).
5. Brenner A, Riddell G E, *J Res. Nat. Bur. Stand.* **37**, 91, 1946.
6. Ibid. **37** (1946) 31–34.
7. Brenner A, Riddell G E, *Proc. Amer. Electroplaters Soc.* **33**, p16, 1946.
8. Brenner A, Riddell G E, *J Res. Nat. Bur. Stand.* **39**, p385–395, 1946.
9. Brenner A, Riddell G E, *Proc. Amer. Electroplaters Soc.* **34**, p156–170, 1947.
10. Gutzeit G, *Plating* **46**, p1158–64, 1959.
11. Ibid. **46**, p1275–78, 1959.
12. Ibid. **46**, p1377–78, 1959.
13. Ibid. **47**, p63–70, 1960.
14. Strafford K N, Datta P K, O'Donnell A K, *Mat. Des.* **3**, 6, p608–14, 1982.
15. Datta P K, Strafford K N, Storey A, O'Donnell A K, *Coat. Surf. Treat. Corros. Wear Resist.* p46–61, 1983.
16. Strafford K N, Datta P K, Googan C. G, "Wear Resistance". Ed. Ellis Horwood Ltd, Chichester UK, 1984.
17. Fields W, *Maschinenmarkt* **89**, p876, 1983.
18. Tuns J, *Maschinenmarkt* **91**, p95, 1985.
19. Matsuoka M, Imaniski S, Hayashi T, *Plating and Surface Finishing* **76**, 11, p54–58, 1989.
20. Parker K, *Plating and Surface Finishing* **74**,2, p60–65, 1987.
21. John S, Shanmugham N V, Shenoi B A, *Metal Finishing* **80**, 4, p47–52, 1982.
22. Mallory G O, Hajdu J B, Eds, Electroless Plating, Fundamentals and Applications Amer. Electrolpat. Surf. Fin. Soc. Orlando, 1990.
23. Lukes R M, *Plating* **51**, p969, 1964.
24. Mallory G O, Lloyd V A, *Plating and Surface Finishing* **72**,11, p52–57, 1985.
25. Riedel W, Electroless Nickel Plating. Finishing Productions Ltd., Stevenage, Herts, UK, 1991.
26. Crotty D, Electroless Nickel '93. Products Finishing. Orlando Airport Paper 3, Nov 1993.
27. McGarian T, ibid Paper 25.
28. Gutzeit G, Krieg A, U.S. Patent 2.658.841, 1953.
29. Gillespie P A, Private correspondence.
30. Konstantinov N, *Z Anorg. Chem.* **60,** p405, 1908.
31. Hansen M, Der Aufbau Der Zweist offlegierungen, Edwards Brothers, Ann Arbor, 1943.

32. Koeneman J, Metcalfe A G, *Trans. Metall. Soc. AIME*, p 571, 1958.
33. Thube M G, Kulkarni S K, Huerta D, Nigavekar Arun S, *Phys. Rev. B* **34**, 10, p6874-6879, 1986.
34. Lu K, Luck R Predel B, *J Non-Cryst. Solids* **156-158**, p589–593, 1993.
35. Matsuoka M, Imaniski S, Hayashi T, *Plating and Surface Finishing* **76**, 11, p54-58, 1989.
36. Kreye H, Muller H H, Petzel T, *Galvanotechnick* **77**, p561–567, p1986.
37. Kreye H, Interfinish '92: Int. Congr. Surf. Finish. Assoc. Braces. Trod. Superficie. Sero Saulo, Brazil, **1**, p146–55, 1992.
38. Riedel W, Electroless Nickel Plating. Finishing Production Ltd., Stevenage, Herts, UK
39. Parker K, *Plating and Surface Finishing* **68**, 12, p71–77, 1981.
40. Duncan R N, Electroless Nickel '93. Products Finishing. Orlando Airport, Paper 27, November 1993.
41. Goldenstein A W, Rostoker W, Schossberger F, Gutzeit G, *J Electrochem Soc.* **104**, 2, p104, 1957.
42. Randin J P, Maie P A, Saurer E, Hinternam H E, *J Electrochem. Soc.* **114**, 5, p442, 1967.
43. Grewal M S, Sastri S A, Alexander B H, *Metal Progress* **91**, p98, 1974.
44. Bagley B G, Turnbull D, *J Appl. Phys.* **39**, p5681, 1968.
45. Rajam K S, Rajagopal I, Rajagopal S R, *Metal Finishing* **88**, p77, 1990.
46. Schenzel H G, Kreye H, *Plating and Surface Finishing* **77**, 10, p50, 1990.
47. Agarwala R C, Ray S *Z, Metallkd.* **83**,3, p199, 1992.
48. Graham A H, Lindsey R W, Read H J, *J. Electrochem. Soc.* **112**, 4, p401, 1965.
49. Kreye H, Muller H H, Petzel T, *Galvanotechnik* **77**, p561, 1986.
50. Mahoney M W, Dynes P J, *Scripta Metallurgia* **19**, p539, 1985.
51. Duncan R N, "Properties of Electroless Nickel Coatings: Effect of Deposit Composition", SUR/FIN '90, American Electroplaters and Surface Finishers Society, Boston, July 1990.
52. Weil R, Lee J H, Kim I, Parker K, *Plating and Surface Finishing* **76**, 2, p62, 1989.
53. Shiosaki M, Sumi K, Japan Kokai 77,150,338 (Cl.C23F1/00) 14 Dec 1977 Appl. 76/68,033 09 Dec 1976 C.A. 89 47742.
54. Kuwajina H, Kamiyama M, Japan Kokai Tokkyo Koho J P 61,170, 584 [86,170.584] (Cl.C23F1/28) 01 Aug 1986, Appl. 85/11,277, 24 Jan 1985 C.A. 106 161074g.
55. Agarwala R C, Ray S, *Z Metallkd* **83**, 3, 1992.
56. Albert P, Kovac Z, Lilienthal H, McGuire T, Nakamura Y, *J Appl. Phys.* **38**, 3, p1258, 1967.
57. US Patent 2.726.170, 1954.
58. Beddingfield P B, PhD Thesis, The University of Northumbria at Newcastle UK, 1993.

Chapter 4

Thermal spraying: an overview

K A Harrison – Metco Ltd

"He that will not apply new remedies must expect new evils, for time is the greatest innovator" – *Francis Bacon.*

4.1 Introduction

Although *thermal spraying* has been in existence for over seventy years, it is only in recent times that the technology has been used seriously as a remedy to combat wear, corrosion, heat, oxidation and other problems occurring across the whole spectrum of the manufacturing and engineering industries.

Thermal spraying is an efficient, cost effective method for enhancing the surface of components to protect against degradation. In the past, thermal spraying, like other surface engineering techniques, has sometimes fallen into disrepute due to failures occurring because of incorrect application. Some engineers will tell of a "bad" experience with "metal spraying". It must be stressed, however, that these prejudices have been overcome and thermal spraying is now widely used, for example, in the manufacture of gas turbines and other engines, often on critical parts. Surgical implants sometimes have thermal spray coatings on them to promote better fixation in the human body. These two examples well illustrate that thermal spraying is an accepted technology in its own right.

In 1910, Schoop [1] patented the deposition of metal by an oxy-acetylene torch process; this was the forerunner of thermal spraying. Development of spraying techniques and equipment progressed slowly during the 20s and 30s. By the time war came, shortage of strategic materials for components led to metal spraying being used to salvage worn parts such as Royal Navy crankshafts and London taxi-cab kingpins. It was not until the late 50s/early 60s that the combustion powder and plasma powder techniques were developed giving a wider choice of processes and materials deposited.

Since then, thermal spraying has grown rapidly giving rise to the *low pressure plasma spraying* and *high velocity oxy fuel* techniques. Thermal spraying has moved forward with the new technology available, i.e. computer aided spraying systems, multi-axis manipulation machines controlled by microprocessors are now standard throughout the industry. Microprocessor monitored, closed-loop control of all key parameters together with statistical analysis of process data is now available, leading to a greater level of process control and repeatability. Thermal spray coatings are now being specified in many critical applications and are also being used not only to produce coatings but to form complex, three dimensional, free-standing shapes of metals, ceramics, and ceramic and metal matrix composites. It is seen as an important process in these emerging technologies.

Figure 4.1 schematically illustrates the principle of thermal spraying; the spray material, in the form of a wire or powder, is precisely fed into a heat source and is given sufficient thermal energy to heat the particle to an advanced plastic state. This near-molten particle is rapidly accelerated to impact onto a prepared substrate. On collision, the particle deforms to form a splat, cools rapidly and adheres to the surface. Subsequent particles build-up to produce a discrete coating. The structure and physical properties of the coatings so formed can vary a great deal and are generally quite different from those of the parent material. Characteristics such as porosity, oxide content, macro- and micro-hardness, tensile strength, adhesion and surface finish will depend on the material and on the spray process used and are further influenced by multi-layer coatings and variations in equipment parameters.

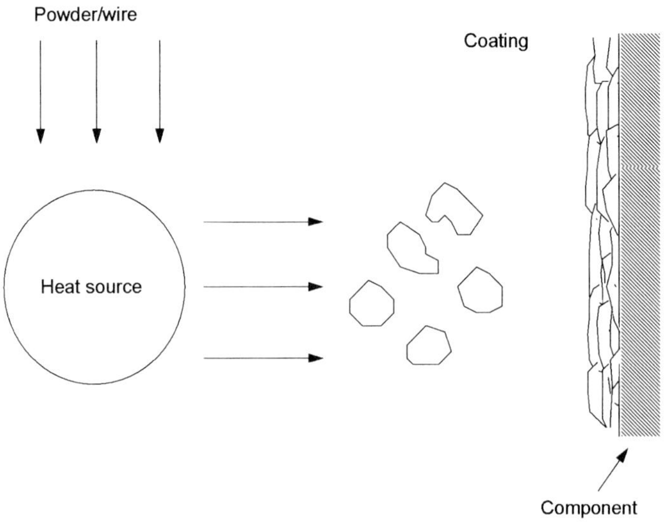

Figure 4.1 Basic principle of thermal spraying.

Thus, thermal spraying is a balance between the thermal and kinetic energy of the particle. The way in which the particle deforms and cools will be an important factor in the resultant coating adhesion and microstructure. The unique nature of thermally sprayed coatings lends itself to a microstructure being tailored to suit a particular design requirement, e.g. the distribution of tungsten carbide crystals in a cobalt matrix to give a wear resistant, relatively ductile surface. It should be noted that normally very little heat is carried over to the base material making thermal spraying a metallurgically cold process. Indeed, with rapid traversing and auxilliary cooling thermally sensitive materials such as polymers and other non-metallics can be successfully coated.

Thermal spraying has grown considerably over the last two decades. This growth has been fuelled by market requirements. New process technology and materials development have evolved to meet the application needs.

4.2 Characteristics and markets

The functions and hence application areas of thermal spray coatings can broadly be defined as follows:-

1. restoration of worn or mismachined dimensions
2. wear resistance
3. oxidation resistance
4. atmospheric corrosion resistance
5. resistance to chemical attack
6. electrical resistance and conduction
7. heat resistance and conduction
8. abradable coatings for clearance control
9. biomedical
10. repair and remanufacture
11. production of free-standing shapes of metals, ceramics or matrix compositics, and
12. fabrication of high temperature superconductors.

The potential market for thermal spray coatings is very large. The estimated size of the US market in 1990 was \$600—700 million; this is expected to increase to \$2000 million by the end of the century [2]. Likewize the estimated European market for 1990 was \$525 million which is expected to rise to \$1,750 million by the year 2000.

Although widely used in industry, thermal spraying is not the panacea to solve all surface engineering problems. There are limitations on the use of thermally sprayed coatings and they can be summarized as:-

a. Thermally sprayed coatings do not add to the mechanical strength of the component; they only restore dimensions or enhance the surface properties.
b. The fatigue strength of the component may be affected by the method of

surface preparation and by the coating. The fatigue strength might be lowered on highly stressed components and must be fully assessed.

c. Some thermal coatings are prone to edge damage and wherever possible, should be contained in an undercut. If this is impractical, then a generous radius should be applied to the component to minimize edge loading. Thermally sprayed coatings will not tolerate high point or line contact loads. However, high squeeze bearing loads in excess of 750MPa can be sustained. Some coatings, especially ceramics, are brittle by nature and have limited ductility and are, therefore, unsuitable for high impact applications and should not be used on components subjected to severe distortion or flexing.
d. There are a small number of coatings based upon nickel–chromium–silicon–boron which are metallurgically fused, after spraying, into the substrate. These dense, highly bonded coatings can withstand a higher degree of loading than normal unfused coatings. However, they have the disadvantage that the component and coating have to be heated to 1050°C to enable the diffusion process to take place. This may cause severe distortion or affect the metallurgical properties of the substrate.
e. Thermal coatings tend to be porous by nature and where this is likely to be a problem in corrosive environments, then the coatings should be suitably sealed with an organic sealant. However, in a lubricated bearing application, the surface may be better left unsealed for the pores to act as lubricant reservoirs.
f. Any previous hard surfacing or surface treatments in the coating area must be removed prior to the application of the thermally sprayed coating and all chemical and metallurgical processing should be completed before the coating processes.

4.3 Materials and processes

Any material that has a well defined melting point and which is not decomposed when heated can be thermally sprayed. Traditionally, metals that could be easily drawn into wires were used as the feedstock for the production of sprayed coatings.

Since the development of the combustion and plasma spraying processes gave the ability to use powders instead of wires, the choice of spray materials was greatly extended. Materials that cannot be drawn into wires can be economically produced as powders and there are over 200 different thermal spray powders available compared with approximately 50 wires.

As the number of spray materials has increased, then the requirements for coating repeatability and quality have also increased. Manual operation is gradually giving way to automated spraying and work handling techniques. Computer aided spraying and micro-processor controlled manipulation are becoming standard within the industry. The development of new powder feeding devices has improved feed rate accuracy and consistency.

Recent advances in powder technology have resulted in the production of fully alloyed, smooth, spherical powders which are free-flowing and have tailored particle size distribution [3]. These characteristics, combined with the development of sophisticated powder feeding devices, have improved spray rate accuracy and consistency leading to reproducible, high quality coatings.

Thermal spray materials can be categorized as follows:-

a. *Iron, nickel and cobalt-based materials*. This group comprizes carbon and stainless steels for wear resistance and corrosion resistance. It also includes exothermic nickel–aluminium materials that have enhanced bond strength to most substrates, and are used as bond coats and as 'one-step' coatings in their own right to combat wear or corrosion. Nickel, nickel chromium, or cobalt-based alloys are used to combat wear and oxidation at high temperatures.
b. *Self-fluxing alloys* based on *nickel–chromium–boron–silicon*. This small family of materials is sprayed metallurgically cold and then subsequently heat-treated at approximately 1050°C to produce dense, pore-free coatings that are metallurgically diffusion bonded to the substrate. They are used to provide high wear resistance and corrosion resistant coatings used under occasionally moderately high loads. Often additions of tungsten carbide cobalt aggregates are used to enhance the wear resistance of these materials. Fused coatings can only be used on substrates that will tolerate the high temperature heat treatments used to fuse the coatings.
c. *Non-ferrous metals*. Copper-based alloys such as the bronzes and brasses are used as soft bearing surfaces. Copper is used for electrical conductivity and for RFI (radio frequency interference) and EMI (electromagnetic interference) shielding. Aluminium is used to produce coatings resistant to atmospheric and chemical corrosion and has excellent heat and electrical conductivity – it is also used for RFI and EMI shielding. Zinc is used to combat atmospheric and fresh or salt water immersion corrosion. Tin is used as a fine, dense coating to protect food vessels, to repair glass-lined tanks and to shield against RFI and EMI.
d. *Oxide ceramics*. Aluminium oxide, titanium dioxide, chromium oxide and mixtures of aluminium oxide and titanium dioxide are used to protect surfaces against hard bearing counter-faces, abrasive grains, hard surfaces, fretting, cavitation and particle erosion. They normally have good corrosion resistance to most acids and alkalines, wetting resistance to aqueous solutions and high dielectric strength. Refractory metal oxides, based on zirconium oxide stabilized with calcium oxide, magnesium oxide, yttrium oxide or cerium oxide are used as thermal barrier coatings in gas turbines and other heat engines.
e. *Tungsten carbide*. Composites of tungsten carbide and cobalt are used to produce very hard, very dense, very wear resistant coatings, particularly resistant to fretting at low temperatures. They are also resistant to wear by abrasive grains, hard surfaces and particle erosion and can be used at temperatures up to 500°C.

f. *Chromium carbide.* Blends of crystalline chromium carbide and nickel–chromium alloy are used to produce smooth, dense, strongly bonded coatings having high temperature wear resistance and oxidation resistance up to 850°C.
g. *Refractory metals.* Refractory metals such as molybdenum are used as anti-scuff coatings on automotive components, for example, piston rings and the bearing surfaces of syncromesh cones. Tungsten is used to produce coatings which have excellent resistance to electric arc erosion. Tantalum is used to repair chemical processing vessels such as those used in the manufacture of dry cell batteries.
h. *Cermet materials.* Mixtures of metals and ceramics are used to form cermet coatings which are used as intermediate coatings in thermal barrier systems to combat thermal mismatch between substrate and refractory oxide ceramic top coating. Cermets are also used as abradable coatings for clearance control purposes in gas turbines.
i. *Miscellaneous materials.* Hydroxyapatite is used to form coatings that help promote bone fixation on surgical implants such as replacement hip prostheses and artificial knees. Thermally sprayed coatings of yttrium, barium, copper oxides and bismuth, calcium, strontium, copper oxides are used to fabricate high temperature superconductors.

There are four basic thermal spray processes:-

a. *Combustion wire spraying* (see Fig. 4.2) – an oxygen/fuel gas flame is produced in a nozzle, the spray material in the form of a wire is precisely fed through the centre of the flame. Compressed air is then used to reduce the near molten tip to a stream of small particles, which are directed on to the prepared substrate.
b. *Electric arc wire spraying* – uses two conductive wires which are melted by resistance heating; compressed air then atomizes the molten metal and projects it as a spray stream. Generally, arc sprayed coatings are more dense and exhibit higher bond strengths than combustion sprayed coatings; and can be sprayed at much higher deposition rates.
c. *Combustion powder spraying* (see Fig. 4.3) – in this LVOF process, powder is precisely fed into a comubustion flame where it is melted and propelled as a spray stream by the inherent velocity of the flame. The velocity of the molten powder particle is sometimes increased by the use of high pressure compressed air.
d. *Plasma spraying* (see Fig. 4.4) – here a plasma-forming gas, argon or nitrogen, is passed between a tungsten cathode and a copper anode. The anode is in the form of a convergent-divergent nozzle. A high voltage arc is struck between the two electrodes and ionizes the gas. As the plasma exits the nozzle, it returns to its natural gaseous state liberating extreme heat. The powder spray material is injected into the hot plasma stream in which it becomes molten and is projected at high velocity onto a prepared substrate. The resultant coatings are strongly bonded and have exceptionally high integrity. When

operating with argon or nitrogen as a primary gas, the arc voltage is a function of the geometrical relationship between the electrode and the nozzle. In order to increase the power level of the gas without prohibitively increasing the current, the electrical properties of the arc are modified by the addition of a secondary gas, either hydrogen or helium [4]. Plasma spraying is usually carried out under normal atmospheric conditions. This is known as *air plasma spraying (APS)*. Plasma spraying can also be carried out under *vacuum (VPS)* or more accurately under low pressure (LPPS). Fig. 4.5 illustrates a typical vacuum plasma spraying system; here a vacuum chamber containing the plasma system is evacuated and back filled with argon to 2.67–13.33 kPa. The coatings produced in this controlled atmosphere have the following advantages over APS:-

- low porosity,
- high density,
- no included oxide,
- minimal chemistry and metallurgy changes between powder and coating,
- thick coating due to lower thermal stresses.

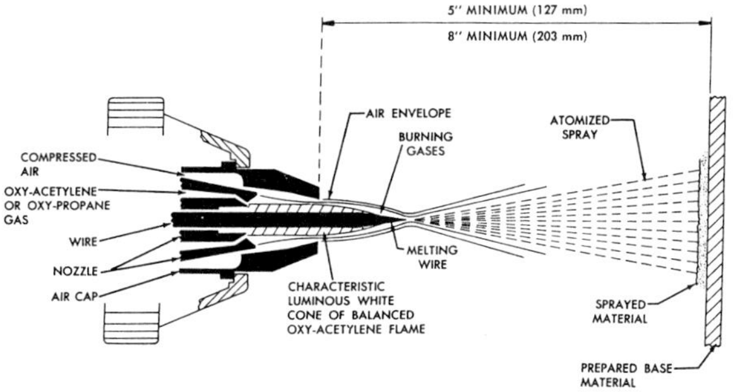

Figure 4.2 Typical combustion wire spray gun.

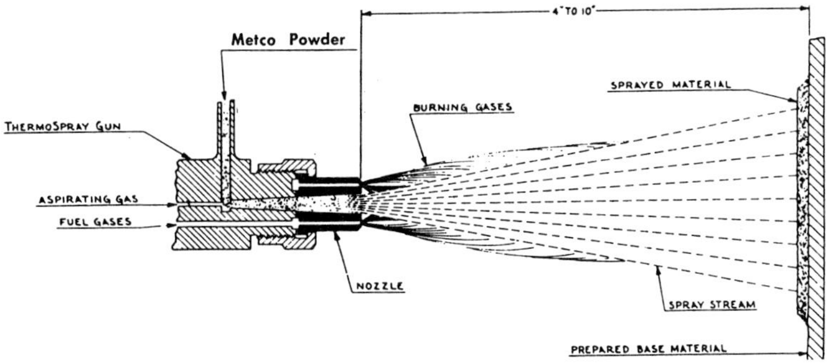

Figure 4.3 Typical 'Thermospray' gun.

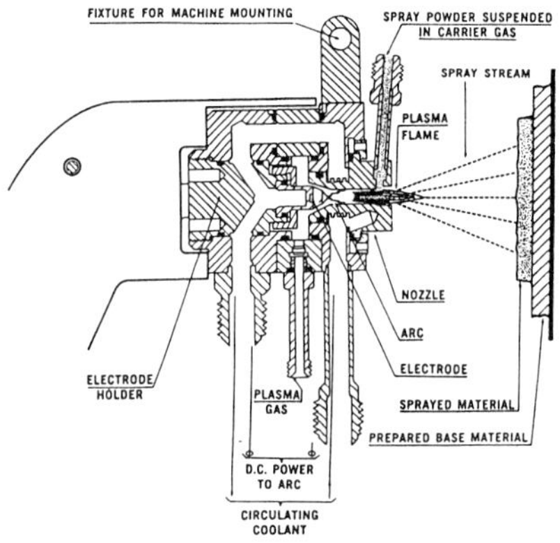

Figure 4.4 Typical plasma spray system.

An alternative to the vacuum plasma route for producing very high integrity coatings is to shroud the plasma with an inert gas and produce coatings almost comparable to the VPS but with restrictions of preheat temperature for components to be sprayed. Plasma spraying is normally a DC non-transferrable arc process, although transferred arc systems are sometimes used. Plasma spraying can also be carried out under water *(UPS)*. *Induction coupled plasma (ICPS)* and *radio frequency plasma (RFPS)* systems have also been used for thermal spraying.

e. *High velocity oxygen fuel process (HVOF)* (see Fig 4.6) – utilizes the energy

Figure 4.5 Illustrating a typical vacuum plasma spraying system.

released by the continuous detonation of oxygen/fuel gas (hydrogen, propane or propylene) in a specially designed spray gun. The spray powder is accurately metered into the spray gun combustion chamber where it is heated and rapidly accelerated up to 800 m/sec. The high kinetic energy and low thermal energy result in coatings with high density, high integrity and very high bond strengths. The low residual thermal stresses associated with this process mean that coatings can be sprayed to thicknesses not normally associated with dense thermal sprayed coatings.

Union Carbide's proprietary D-Gun process is a pulsed detonation type rather than continuous. Oxygen/acetylene mixtures are ignited up to 8 times per second in a tube, the spray powder is metered into the tube and each detonation produces a circular deposit approximately 25mm in diameter. The coating is built-up by a series of overlapping deposits to the required thickness [5].

4.4 Sprayed coatings

The metallurgy of the thermally sprayed coating is normally different from the wrought or cast material, e.g. metal coatings tend to be harder than the same metal in wrought or cast form, making the sprayed metal coating better suited for wear resistant applications. The main reason for the metallurgical difference is the rapid quenching of the molten sprayed particles as they strike the substrate and are flattened into elongated platelets. The resultant microstructure will show a degree of microporosity, a small percentage of unmolten particles and some included oxide. A typical thermally sprayed coating microstructure is shown in Fig. 4.7. The coating structure is not only a function of the spray material but also of the process and process parameters used. By optimizing the spray powder morphology, size and size distribution and by choosing the correct combination of spray process and spray parameters, the coating microstructure can be tailored

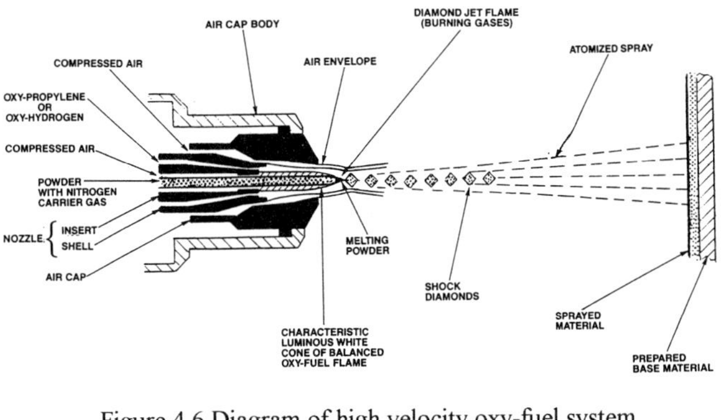

Figure 4.6 Diagram of high velocity oxy-fuel system.

Figure 4.7 Thermally sprayed coating microstructure.

to satisfy the needs of a specific application; in many cases, dense coatings with the minimum of included oxide can be produced. Figure 4.8 shows an optimized tungsten carbide/cobalt coating with minimal porosity and included oxide; here the even distribution of the carbide phase can easily be seen. The physical properties of the coating are very much dependent upon its microstructure. Bond strength and thickness limitation are also influenced by the build-up of residual stresses during the deposition process [6].

Figure 4.8 An optimized tungsten carbide/cobalt coating with minimal porosity and included oxide.

The main bonding mechanism was thought to be mechanical interlocking of the splat on to the substrate surface asperities. It is now considered that the adhesion of sprayed coatings derives from a mixture of chemical and physical forces dominated by a chemical mechanism when the fast moving near-molten particle interacts with the prepared substrate. In order to achieve maximum bond strength the surface preparation of the component is of paramount importance. The surface to be coated must be absolutely chemically clean, free of any oils, greases, waxes or other kinds of contamination such as scales and oxides. Normally, pre-cleaning is carried out by a degreasing fluid. The surface activation is achieved using a grit-blasting process utilising aluminium oxide or chilled iron grit. The purpose of this process is to remove any residual microscopic contaminant and to increase the surface area. Ideally, the coating should be deposited as soon as possible after surface activation in order to avoid any substrate surface deterioration.

The majority of the thermally sprayed coatings theoretically have no thickness limitation. Practically, there are reasonable limits that vary depending on the application, the thermal spray method used and the material sprayed, some of the more highly stressed coatings such as tungsten carbide, chromium carbide and oxide ceramics such as aluminium oxide are normally limited to thicknesses of 0.4mm. As coating thickness is directly proportional to cost, then most users keep coating thicknesses to a minimum. In appearance, most as-sprayed thermal spray coatings have a dull, matt finish. Depending on the material application method and spray parameters, the as-sprayed surface finish may range from as low as 25μm AA for HVOF coatings up to 750μm AA and above, the high profile coatings are applied by electric arc spray. In general, thermal spray coatings can be machined to 10–20μm AA and can be ground to 2.5–5μm AA. Many plasma coatings can be lapped to better than 0.5μm AA.

Sprayed coatings can be very hard and wear resistant, e.g. plasma sprayed tungsten carbide has a macro-hardness of RC75 and a micro-hardness of 800–1200 VHN. They can also be soft enough to allow the embedding of abrasive particles and to permit deformation to accommodate misalignment of bearing surfaces; examples of this are combustion wire sprayed aluminium bronze and babbitt.

Both zinc and aluminium coatings are used for corrosion prevention. Since both are anodic to steel, they will corrode preferentially to the steel. To maximize the electrolytic protection of thermally sprayed zinc or aluminium, coatings can be enhanced by applying a primer that acts as an inhibitor and by applying one or more seal coats to provide an impervious barrier film.

In order to increase coating consistency and repeatability and the increasing use of thermally sprayed coatings to solve complex problems then a more thorough knowledge of the structural behaviour and residual stress distribution of coatings is required. This has lead to an increasing use of thermal mechanical analyses and modelling techniques to give a greater understanding of coating structures, behaviour and quality assurance [7].

There is an increasing use of statistical techniques in optimizing the spray parameters and in understanding the relationship between spray parameters and coating properties. Statistical process control techniques are being used to improve the quality and reduce the rejection rate during the manufacture of powders and spray equipment.

4.5 Coating quality and repeatability

In order to achieve consistent, high quality coatings to meet the more demanding performance requirements of today's applications, there is a need to control every stage of the coating's production. As the only true way to evaluate fully a coating is to use a destructive metallographic technique, then the emphasis must be on process control. Good control of the process must be carried out at every stage.

4.5.1 Substrate activation

This is carried out only on chemically clean surfaces by grit blasting. The grit size and chemical composition should be specified. The grit must be kept clean and dry and should be regularly monitored. The parameters for activation, i.e. nozzle size, nozzle-to-component distance and air pressure, must be specified and controlled. Substrate activation is carried out only on a chemically clean surface.

4.5.2 Spray material

All thermal spraying materials are manufactured to rigorous quality control standards necessary to ensure high coating quality and repeatability. Wire, of carefully controlled chemical composite, is drawn to a consistent diameter in one piece, free of any slivers or other defects which might impede consistent feeding. Likewize powder manufacture is carefully controlled to ensure chemical composition, particle shape, particle size and size distribution. Care is taken to avoid contamination during the manufacturing process.

4.5.3 Coating deposition technique

The critical factors that need to be controlled and monitored during deposition of the coating are:

- gun-to-workpiece-distance,
- gun to workpiece angle,
- gun/workpiece relative speeds,
- parameter control,
- thickness control, and
- powder feed rate.

Thermal spray systems can be manually set up to a specified technique sheet.

Simple rotation of a cylindrical component together with horizontal or vertical traverse of the spray gun will give a certain degree of coating repeatability. However, to achieve total coating reproducibility and repeatability, there is an obvious need to automate the process wherever possible. Automation allows all the process variables to be monitored and precisely controlled. All of the thermal spray processes can be run by closed-loop computerized systems which will actively monitor and control all process parameters which have an influence on the properties of the deposited coating. This ensures high coating quality, part-to-part and batch-to-batch coating uniformity. These computerized systems can be used in conjunction with automated gun handling and workpiece manipulation equipment. Specifically designed robots are now available for use with thermal spray equipment allowing repeatable spraying of large numbers of components with different shapes. These robots are typically 6-axes machines with heavy-duty, robust, electric drive DC servo-motors and have been specifically designed to operate in dust-laden atmospheres. In addition, they are shielded to eliminate any interference effects generated by the spray systems. The robots are easily programmable to give a wide range of traversing patterns. The robots can also be used in conjunction with a CAD/CAM system and can be incorporated with a microprocessor-controlled, multi-axis workpiece manipulator, thus complete integrated control of the spray system, robot and workpiece manipulator can be achieved. Figure 4.9 shows a typical robotic installation.

New powder feed devices based either on the fluidized bed principle or volumetric rotating plate principle allow accurate metering of the spray powder into the heat source. These devices can be used in conjunction with the computerized gun systems and can incorporate electronic mass flow meters and microprocessor control. The powder feed rates are controlled within 2% thus achieving the optimum heat energy balance between powder particle and heat source to give consistently high quality coatings with high deposit efficiencies, high integrities, low oxide inclusions and with the minimum of unmelted particles.

4.6 Quality assurance

Reliable coatings are only achieved by good process control at every stage of the processing cycle. Process variability can be minimized by computer monitoring and control and by automated spray techniques. Quality should be considered from coating system concept to the finished product.

Process control encompasses:-

- spray material – conformance to specification
- substrate – free of defects, contamination, etc.
- surface preparation
- spraying parameters
- spraying technique

Figure 4.9 A typical robotic installation.

- coating thickness
- finishing
- sealing
- post-spray treatments.

Destructive coating integrity tests should be carried out only on test pieces or dummy components, and may comprize:-

- adhesion
- hardness – micro and macro
- microstructure – porosity, included oxide, unmelted particles
- abradability (where applicable)
- thermal shock
- thermal cycling
- thermal conductivity
- resistivity
- thermal expansion
- Young's modulus
- tensile bond strength.

Optimum coating integrity for a specific application can be achieved by evaluating dummy or scrap components under rig or actual service conditions. Even though the only true test of the coating is to evaluate it destructively, there has been a considerable effort put into the development of non- destructive testing techniques, e.g.

1. die penetrant
2. magnetic particle inspection
3. dimensional inspection
4. visual inspection at low magnification for surface defects
5. ultrasonic and acoustic emission techniques [8]
6. laser wave and holographic interferometry techniques [9]

7. pulsed video and other thermograph techniques [10].

There is no universal NDE test available but some of the above techniques have been used successfully on a number of specific applications.

It has been shown that reproducible and reliable coatings, even from process-sensitive powders, can be readily achieved by careful control of the process variables.

4.7 Applications

The market for thermally sprayed coatings is very large and also very diverse. Applications can be found across the spectrum of engineering and manufacturing, from bridges to bone fixation, from gas turbines to golf clubs.

Thermal spraying allows the designer complete freedom of choice of base materials and subsequent surface coatings. Qualities of a material best suited to service conditions are combined with a base material having other attributes such as lightness, strength and low cost. This allows the designer to value engineer high performance parts by applying premium materials over lightweight or lower cost substrates.

Applications continue to be developed where the potential of thermal spray coating characteristics are at an advantage. Large surface areas and complex shapes can be coated with tailored or composite microstructures to suit particular service needs. Free standing shapes of metals, ceramics, composites or polymeric materials can be produced by thermal spraying. These structures have properties that are difficult to obtain by other techniques and may offer normal solutions to difficult manufacturing problems.

The main areas where thermally sprayed coatings are used can be summarized as follows:-

4.7.1 Aerospace

Thermally sprayed coatings are extensively used in the manufacture and repair/overhaul of aero engines, and may be grouped as:-

a. *Wear resistance*. Coatings such as tungsten carbide/cobalt, copper–nickel–indium or chromium carbide/nickel–chromium are used to combat surface fatigue problems such as fretting of mating surfaces due to vibration and temperature changes. This includes combating hammer-wear on the compressor blade, mid-span support faces or protecting the abutment faces of turbine blades.
b. *Clearance control*. To minimize air leakage around the tips of blades or through labyrinth seals, abradable coatings are used, which can be "machined" by the rotating component. These allow the designer to specify theoretically "zero clearance" for the operating range of engine speeds. Abradable coatings include

silicon–aluminium/polyester, nickel–chromium/boron nitride and nickel–graphite materials.

c. *Thermal barrier*. Thermal barrier coatings (TBCs) such as yttrium oxide stabilized zirconium oxide or magnesium oxide stabilized zirconium oxide are used on components such as combustion chambers, combustion cans, nozzle guide vanes and turbine blade platforms. TBCs protect the component against thermal degradation and reduce the rate of heat transfer through the component thus minimizing cooling requirements.

d. *Oxidation resistance*. High temperature alloys of the MCrAlY family are used to protect the gas-washed surfaces of turbine blades and other hot-end components against long-term oxidation in the jet stream. These coatings are normally applied by the low pressure plasma spraying system.

e. *Salvage/repair/re-manufacture*. Production parts or casings mis-machined undersize are brought back to drawing specification by applying a thermally sprayed coating. Service worn components can be rebuilt, by thermal spraying, faster and at lower cost than by welding, plating or sleeving. The repair costs much less than buying a replacement part and the thermally sprayed, coated part often lasts longer than the original. Remanufactured components can often be given an "as new" guarantee.

As well as aero engine applications, which have been well documented [11], thermal spraying is used on air-frame components, e.g. aluminium bronze coatings are used to reclaim the bores of landing gear housings which have suffered from galvanic corrosion. To prevent further corrosion occurring, the coatings are sealed with either an air-drying phenolic resin or glyceride sealant. The sliding surfaces of aerodynamic control surfaces, such as slat or flaptracks are reclaimed using a tungsten carbide–cobalt coating and sometimes an intermediate coating of similar hardness to the substrate is used to build up the dimension. Composite aircraft panels can be metallized with aluminium to protect against lightning strikes or to dissipate static electricity. Some panels have been coated with refractory ceramics to protect them against the heat of missile exhausts. Plasma spraying has been used to coat or even fabricate venturi nozzles for rocket applications.

As aerospace designers strive to make their products more efficient, the use of thermally sprayed coatings will increase.

4.7.2 Heat engines

Industrial gas turbines are also extensive users of thermally sprayed coatings. The application areas are the same as those in aero engines.

There is a predicted increase in the use of thermal coatings in automotive applications. Research and development programmes are under way evaluating refractory ceramic coatings on components such as piston crowns, cylinder heads and valve faces, i.e. combustion chamber components.

The outside diameters of top piston rings and the bearing diameters of syncromesh cones in reciprocating engines are coated with molybdenum or molybdenum alloy which provide excellent scuff resistance in minimally lubricated environments. Ceramic coatings of aluminium oxide are already being used to insulate some components involved in electronic engine management systems and plasma sprayed zirconium oxide is being used as an oxygen sensor material.

4.7.3 Power generation

Coal-fired boilers that burn high sulphur, high chlorine coal suffer from fireside-corrosion. This phenomenon can cause water carrying tubes in the waterwall, re-heater and superheater areas to lose thickness and eventually leak. These leaks can lead to premature shut-down of the boiler unit causing very high generation losses. Plasma sprayed 50 nickel/50 chromium coatings can reduce corrosion rates by up to 40 times thus allowing units with severe corrosion problems to last between planned maintenance periods. Plasma spraying is carried out on new tubing under factory conditions or can be done in-situ on corroded tubes that have reasonable residual wall thickness. Plasma spraying is approved by the UK and US power generation companies for combating fireside corrosion.

In addition, coatings of tungsten carbide/cobalt are used to increase the service life of axial fan blades that suffer from erosion. Bearing journals and seal diameters of general machinery used in power generation are often reclaimed by coating with a high chromium stainless steel.

4.7.4 Steel industry

The aggressive environment of a steel mill where red-hot strip, slab or billet is conveyed from one area to another can cause extreme wear and abrasion to the conveyor and guide rolls. This is due to heat and mill scale produced under working conditions, water is sometimes used as a coolant which may cause corrosion problems. The working faces of table guide rolls, conveyor rolls, deflector rolls and pinch rolls are hardfaced with a nickel–chromium–silicon–boron fused coating which can increase their life by up to a factor of six. Table guide rolls, which sometimes fail prematurely causing unplanned shutdown of a production line, can be coated to give a predictable service life thus enabling planned maintenance to be carried out.

A further example is nickel aluminide coatings on basic oxygen furnace fume hood cooling tubes which increase their life by up to 3 times by protecting against abrasive iron oxide at high temperatures. Pure aluminium is sprayed on to hot steel billet or slab through stencils for identification purposes. The aluminium fuses to the hot metal giving a permanent, highly visible marking even if the steel is reheated during rework to a red-hot state.

4.7.5 Petro-chemical industry

Vacuum fused coatings of nickel–chromium–silicon–boron and non-fused HVOF tungsten carbide/nickel–cobalt–chromium provide wear resistance and corrosion resistant surfaces on gate valves used in critical sub-sea applications.

Thermally sprayed coatings of chromium oxide or aluminium oxide are used to protect pump sleeves and impellers against corrosion and wear in the chemical industry; tantalum coatings are used to protect processing vessels in dry cell manufacture.

4.7.6 Textiles

The abrasive nature of synthetic fibres creates a major wear problem on textile machine components such as thread guides, draw pins and draw rolls.

The application of ceramic coatings such as chromium oxide, aluminium oxide or aluminium oxide/titanium dioxide can minimize these problems significantly. These coatings can be left as-sprayed or finished in a specific way to impart the necessary 'slip' or 'grip' to the thread.

4.7.7 Paper and printing industries

Bearing journals and seal diameters of rolls used in the paper making and printing industries are coated with 12% chromium steel to reclaim worn or scored dimensions, the coated surfaces often giving a longer life than the base material.

The working faces of the very large *Yankee drier rolls* used in paper manufacture are re-lifed with a plasma sprayed stainless steel coating. The coating process can take many hours to build-up the required thickness and is often performed in-situ to avoid the expensive removal and refitting operations.

Calendar rolls which are used for polishing paper and which suffer from severe abrasion are coated with HVOF tungsten carbide cobalt coatings and are produced to very high surface finishes.

Anilox rolls transfer the ink from the pick-up roll to the print roll in the flexographic printing process. These rolls are protected against severe abrasion and sometimes corrosion by coating with plasma sprayed chomium oxide. After coating, the rolls are ground to a mirror finish and then laser engraved to produce the correct ink cell configuration, which can be in excess of 400 cells per linear inch. In order to prevent malformation of cells, it is important to eliminate any metallic contamination in the coating. The spray powder has to be free of any chromium particles and special nozzles are required to avoid metallic spatter.

4.7.8 Marine industry

Ship hulls and structures are coated with zinc or aluminium to protect against sea water corrosion. Many components previously cadmium plated are now coated with zinc or aluminium, thus eliminating the ecological problems associated with the disposal of cadmium plating effluent.

Non-skid walkways, ladders and helicopter landing pads are produced by electric arc spray aluminium using special parameters which give a high profile surface finish.

Service worn auxiliary diesel engine components, such as crankshafts, pistons, connecting rods and valves, are reclaimed by plasma spraying. Propeller blades, which suffer from cavitational and/or silt erosion, are protected by a plasma sprayed aluminium oxide/titanium dioxide coating. This can increase their life by up to 4 times. Bearing and seal diameters of tailshafts, stabilizer shafts and rudder stocks can also be reclaimed by plasma spraying. Bearing houses can be reclaimed by coating with white metal (babbitt) or phosphor–bronze. Armature shaft seal diameters have been reclaimed using tungsten carbide/cobalt coatings. Plasma spray coatings of iron–aluminium–molybdenum alloys have been used to restore compressor crankshaft bearing diameters.

A number of generic commercial shipping repair applications by thermal spraying have been approved by *Lloyd's Register.*

4.7.9 Metals handling and heat treatment

Metals handling equipment, such as crucibles, pouring troughs, thermocouple sheaths and ingot moulds, are protected with aluminium oxide or aluminium oxide/titanium dioxide coatings to combat the aggressive conditions associated with molten metals.

Heat treatment jigs and fixtures, graphite trays and annealing and normalising rolls are coated with stabilized zirconia coatings to protect against thermal degradation.

4.7.10 Electronics

Thermally sprayed coatings have been in use for many years in the electronics industry; paper and polyester wound electrolytic capacitors are metallized with tin/zinc for electrical connection. Similarly, carbon resistors are metallized with copper. Structural polyurethane foam mouldings, such as VDU and computer casings, are coated with zinc, tin–zinc or copper to protect against stray electromagnetic and radio-frequency interference.

More recently, coatings have been used to fabricate power electric substrates [12], both the dielectric and conductive tracks are applied by thermal spraying, and to make electrical contact on to high temperature superconductor substrates [13]. Thermal spraying techniques are also being used to fabricate high temperature superconductors such as the yttrium, barium, copper, oxygen and bismuth, strontium, calcium, copper, oxygen types, especially on to large surface areas or complex shapes.

4.7.11 Biomedical

Porous or high profile coatings of aluminium oxide, titanium–aluminium–vanadium alloy or cobalt–chromium alloys have been used on non-cemented surgical implants to help promote fixation in the human body by allowing skeletal in-growth to occur into or on to the coated surface.

More recently, dense coatings of synthetic bone – hydroxyapatite – have been plasma sprayed on to tooth and orthopaedic implants to give a more rapid and stronger adherence to new bone, possibly leading to "permanent" fixation. These coatings show good biocompatibility and strength [14].

Plasma sprayed fluorapatite coatings are also being considered for biomedical applications [15].

4.7.12 General

All components that suffer from wear, corrosion or thermal degradation are potential applications for thermal spraying. Bearing journals and seal diameters of rolls, cylinders and shafts used in abrasive conditions are ideal candidates for a thermally sprayed coating. Extruders used in the plastics industry are usually hard-faced with a nickel–chromium–boron–silicon alloy to combat wear and sometimes corrosion. Knives and other cutting edges can be coated with a thin tungsten carbide cobalt film to increase their life.

Carbon fibre reinforced composite materials can be metallized to improve their erosion resistance or can be coated with a thermal barrier coating to protect against extreme heat [16].

Thermal sprayed aluminium or silicon–aluminium has been used to produce a porous enhanced boiling surface for cryogenic liquids. Such coatings are shown to have excellent heat transfer capabilities, increasing the heat transfer coefficient by up to 10 times that of smooth plates [17].

Thermal spray techniques are being used to fabricate metal matrix and ceramic matrix composites [18]. Thermally sprayed coatings are also used to enhance the surface properties of a metal matrix composite, e.g. a thin ceramic coating on a metal matrix composite brake disc greatly improves the performance of the disc during racing.

4.8 Summary

The use of a thermally sprayed coating to improve the surface properties of a component is widespread and new application areas are being developed, often for use in critical environments. In order to satisfy these more stringent requirements, a more thorough understanding of the way that the spray material reacts with the heat source and how each molten particle/heat source reacts with the substrate is needed.

More research into these areas is being carried out. The use of modelling techniques is becoming more widespread and statistical process control to evaluate powders and coatings [19] is being applied by users of thermal spray equipment. Spray powder properties have to be carefully controlled to produce optimum coatings for maximum performance in service.

Thermal spraying is playing an increasing rôle not only in salvage of service-worn or mis-machined parts but also in original equipment manufacture and is now widely recognized as an important surface engineering technique.

Acknowledgement

The author wishes to thank Sulzer Metco UK Ltd for permission to publish this chapter.

References

1. Schoop M, patent numbers:- GB 5712; France 406,387; DRP 2,585,005; Sweden 49,270.
2. Gorham Advanced Materials Institute – Report on the North American Thermal Spray Market, Press Release 1990.
3. Dorfman M R et al, "Spherical Ceramic Powders", Proc. 11th International Thermal Spray Conference Montreal 1986.
4. Roumilhac Ph et al "Comparison of Ar-H_2 and Hr-He plasma jet produced by DC plasma torch working either in air or in a controlled chamber at atmospheric pressure", Proc. 12th International Thermal Spray Conference, London, **2**, p61-69, 1989.
5. Tucker Dr RC and H Nitta, "Detonation Gun Coatings – Coating Characteristics and Applications", Proc. ATTAC '88, Osaka, p85–93, May 1988.
6. Rickerby D S et al, "Analysis of the residual stresses in plasma sprayed coatings" 1st Plasma Technik Symposium, Lucerne, **2**, p267–276 1988.
7. Smith R W and Norak R, "Advances and applications in US Thermal Spray Technology -I Technology and Materials", *Powder Technology International,* **23**,3, p147–155, June 1991.
8. Suga Y et al, "Study on the Ultrasonic Test for evaluating the adhesion of sprayed coatings to a substrate", Proc. International Thermal Spray Conference, Orlando, FL, p247–252, 1992.
9. Marks J D et al, "Developments in thermal wave non-destructive testing systems for spray coatings", Proc. 12th International Thermal Spray Conference London, **1**, p49–54, 1989.
10. Milne J M and Scott K T, "Thermal pulse video thermography of sprayed coatings – an inspection technique for bond quality", Proc. 3rd National Thermal Spray Conference, Long Beach, California, p343–380, 1990.

11. Rhys-Jones T N, "Applications of thermally sprayed coating systems in aero-engines, Proc. 12th International Thermal Spray Conference London 1989 1. p87–99.
12. Bjorncklett A et al, "Coating system for electrical applications", Proc. International Thermal Spray Conference, Orlando, FL, p829–834, 1992.
13. Ashworth S P et al, "Electrical contacts to YBCO using metal spray techniques", *Cryogenics*, **29**, December 1989.
14. Berndt C C et al, "Thermal Spraying for bioceramic applications", *Materials Forum*, **14**, p161–1731, 1990.
15. Wolke J G C et al, "Plasma-sprayed fluorapatite coatings for biomedical applications", Proc. International Thermal Spray Conference, Orlando, FL, p471-476, 1992.
16. Ducos M P, "Atmosphere and temperature control plasma spraying thick coatings onto composite resin fibres", Proc 12th International Thermal Spray Conference, Orlando, FL, p121–131, 1992.
17. Ashworth S P et al, "The effect of coating thickness and material on a porous enhanced boiling surface in cryogenic liquids", Proc. Low Temperature Engineering Conference, Southampton 1990.
18. Steffens H D et al, "Production of metal-matrix composites by thermal spraying", Proc. 12th International Thermal Spray Conference, London, **1**, p325–334, 1989.
19. Dorfman M and Debarro J, "Evaluation of Zirconia based powders and coatings using statistical processing techniques", Proc. International Thermal Spray Conference, Orlando, FL, p439–446, 1992.

Chapter 5

Aqueous corrosion: an overview

D Kirkwood – Glasgow Caledonian University

5.1 Introduction

The processes of decay of materials in engineering service have become of progressively greater importance with time, as the following engineering developments have required:

a. The use of materials with greater mechanical properties, through poorer corrosion properties.
b. Slender design philosophies with tighter materials design margins.
c. The use of high value/cost alloys for specialist engineering service.

Regrettably, even today when corrosion costs to the UK economy are running at over £10^9 *per annum*, corrosion processes are still not considered a high priority in the overall engineering design process in many instances. The consequences of this are legion, and it is a matter for us all to ensure that we apply to the fullest extent, existing knowledge and techniques of good corrosion control (*which incidentally will save more money than any other action heading*) [1,2].

5.2 What is corrosion?

There are many ways in which engineering materials can degrade in service – corrosion, erosion, radiation conversion etc. – though there is little doubt that corrosion is the most important, sometimes insidious and widely active phenomenon across the engineering spectrum.

Corrosion may be defined as an unintentional (less commonly intentional) *attack on a material through reaction with a surrounding medium which results in material conversion or loss*. It is important to remember that '*intentional corrosion*' – where we want to encourage efficient corrosion processes – is very important in battery technology, *sacrificial anodes* (zinc and copper), magnesium time-release couplings, and so on.

In the simplest view, corrosion involves a loss of, or conversion from, the metallic state, commonly in reaction with oxygen, sulphur, chlorine sulphate or nitrate ions particularly to give a *corrosion product*. Many corrosion products are not in themselves protective against further corrosion, either because they are soluble, unstable or incoherent in form. Some corrosion products such as oxides of aluminium, chromium and titanium can provide properties of corrosion resistance thereby reducing further loss. This is a simple form of *passivation*, though this must be carefully maintained and preserved otherwise further (and more serious) corrosion problems may result.

Corrosion can take place at high or low temperatures, and in solid, liquid or gaseous media. Generally, the kinetics of corrosion reactions are such that at temperatures lower than 200K (–73°C) the rates become insignificant. So *temperature*, if we can control it, can be a simple and effective form of corrosion control.

Solid-state corrosion is generally a sluggish process, and can be a problem in electronics and battery technology sectors. *Liquid and gaseous media corrosion* are far and away the most common, and unfortunately, they can produce the most aggressive environments.

Taking matters a little further, in terms of materials and media, low/medium-alloy steels remain the dominant engineering materials, for structural and many process applications. These steels have poor intrinsic corrosion resistance and so protection schemes of one kind or another are generally required. In terms of media, water or water-based corrosion is the dominant mechanism. Consequently, the most common corrosion reaction in engineering can be expressed thus:

$$\text{iron} + \text{oxygen (or hydrogen)} \longrightarrow \text{iron oxide (or hydroxide)} \qquad [5.1]$$

The iron oxide or hydroxide formed is manifested by the familiar *rusts* formed on irons and steels.

5.2.1 The corrosion process in metals and alloys

The metallic structure is unique among engineering material structures. For maximum stability, metallic atoms are able to form extremely close-packed crystalline lattices in some cubic variants – simple, body centred cubic (bcc), face centred (fcc) – or hexagonal close-packed (hcp) array. In doing, they must release their outermost valence electrons which migrate freely within the lattice and are not assigned to any discrete atomic bond. They have a "*roving commission*" and are largely responsible for varying properties of metal viz:

- high thermal conductivity
- high electrical conductivity
- unique metallic lustre
- thermionic emission and so on.

In reality of course, engineering metals and alloys are polycrystalline, i.e. there

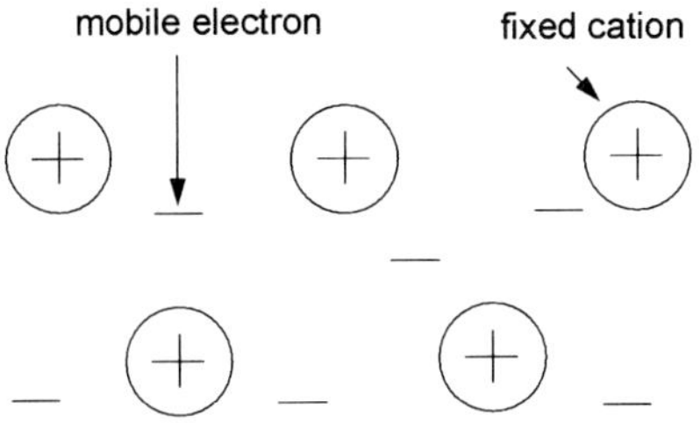

Figure 5.1 A simple view of the metallic structure.

are countless "building bricks" of crystals or grains (as shown in Fig. 5.1) stacked together to form the metal or alloy, hence the performance of such a metal or alloy is often influenced by the structure as a whole. The boundaries of grains are high energy areas due to high internal stresses arising from concentrated dislocation networks. Furthermore some grains may be more or less active than others for a number of reasons:

- differing concentration of solute atoms,
- higher or lower lattice energies due to cold work, heat treatment etc.

In other words there are several mechanisms possible which create energy differences in what might initially appear to be a uniform material. Again, corrosion processes involving metals generally involve electron transfer in some form as well as chemical change – this is called *electrochemical reaction*. Every metal and alloy has a different overall activity level (i.e. tendency to undergo electrochemical reaction) so again we find a further difference in intrinsic activity level or tendency to corrode.

5.2.2 Anodes and cathodes in metal alloys

Arising from the above, let us examine the case of a polycrystalline metal immersed in, say, a low pH aqueous solution.

We see the formation of areas where *cationic dissolution* is favoured (due to relatively *high cationic activity*) and areas where *electron consumption* must be undertaken to allow the corrosion process to proceed (areas of relatively *low cationic activity*). The former areas are called *anodic* sites and the latter *cathodic* sites, and, of course, in any metal we must have both sites co-existing, though they must change in location periodically due to build-up of corrosion product etc. In short at *anodic* sites the following reaction summarises the position:

$$M \longrightarrow M^{Z+} + Z.e^{-} \qquad [5.2]$$

The valence electrons so produced migrate within the metal to *cathodic* surfaces where the two following reactions are possible:

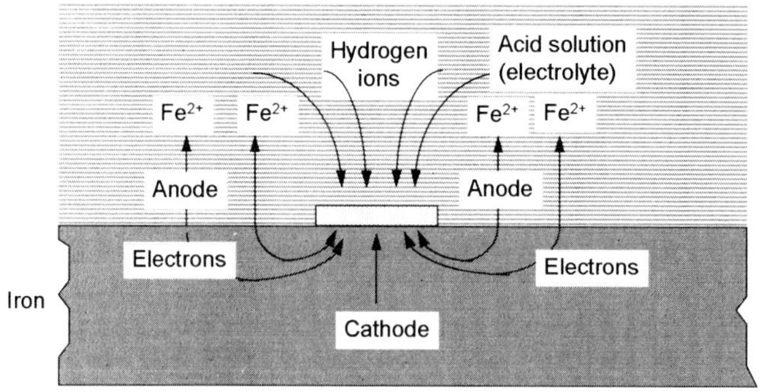

Figure 5.2 Ionic and electron transfer in low pH aqueous corrosion.

$$O_2 + 2H_2O + 4e^- \longrightarrow 4OH^- \qquad [5.3]$$

$$2H^+ + 2e^- \longrightarrow H_2 \qquad [5.4]$$

Consequently, corrosion can only take place at anodic sites and is impossible at cathodic sites. Remember too that anodic and cathodic reactions must be simultaneously active to allow corrosion processes to proceed. From equations [5.2], [5.3] and [5.4] we see that the important parameters in the simple corrosion process are:

- the intrinsic activity or basic reactivity of metal atoms;
- the availability of oxygen or hydrogen ions to participate in the *cathodic reactions*, equations [5.3] and [5.4] and, for oxidation reactions at anodic sites;
- the availability of water for *cathodic reaction*.

5.2.3 The cathodic reactions

In normal sea water for example (pH ~ 7.8 and at normal electrode potentials) oxygen reduction is the most common cathodic reaction (equation [5.3]). This is confirmed for iron by examining the *Pourbaix* diagram for iron and water (Fig. 5.3) and we also find that for most potential variations, in sea water, the hydrogen cathodic reaction becomes very likely indeed:

- in anoxic or stagnant sea water conditions,
- at elevated sea water temperatures (oxygen solubility is reduced),
- where pH is reduced, or
- where excessive cathodic protection is applied (overprotection).

As will be appreciated, hydrogen gas is undesirable though sometimes an unavoidable cathodic product as it can produce *hydrogen embrittlement* in high strength steels and weldments, and give production and safety problems in the internals of tubulars and oil and gas process plant.

5.3 The intrinsic activity of metals and alloys – the galvanic series

The intrinsic activity of metals and alloys is, of course, the basic driving force for corrosion processes to occur. We measure this force in terms of the *electrical pressure* or tendency to produce electrons arising from *anodic reactions* (equation [5.2]). The electrical pressure is more correctly termed the *electrode potential* and is commonly quoted with reference to the *standard hydrogen electrode (SHE)* which is an extremely stable reference electrode. In practice both in the field and in sea water, the *SHE* is not a very practical, nor rugged arrangement and several other reference electrode systems have been developed. These systems with their main uses are summarised in Table 5.1.

In corrosion engineering, *rest potential measurements* are of some value being used in corrosion monitoring applications and in compiling the *galvanic series of metals and alloys*. In the *galvanic series*, metals and alloys are listed according to their intrinsic activity (electrode potential) from the most active members with the most negative electrode potential to the most noble members with the most positive electrode potential. Consequently, the former members are relatively keen to corrode, the latter least known to corrode. A typical *galvanic series* of metals and alloys in sea water is given in Table 5.2.

5.4 Thermodynamics and kinetics of corrosion reactions

5.4.1 The thermodynamics of corrosion – the "Pourbaix diagram"

The various reactions which can occur between a metal and water have been measured, collected and calculated by *Pourbaix* amongst others. These have been combined with solubility data on oxides and hydroxides and equilibrium constants to produce what are effectively phase diagrams (*Pourbaix diagrams*) for metals and water, showing which phases are stable in terms of the important parameters – *electrode potential* (E_H) and *hydrogen ion concentration* (pH).

Table 5.1 Reference electrodes for use in corrosion potential measurement

Reference electrode	Potential (Volts) wrt SHE	Uses
Saturated calomel (SCE)	+ 0.242	General reference electrode
Normal calomel	+ 0.280	Specialist laboratory
Tenth normal calomel	+ 0.334	Reference electrodes
Silver/siver chloride (Ag/AgC1)	+ 0.222	Good rugged seawater reference electrodes
Zinc	– 0.762	Cheap rugged seawater reference electrode
Copper/saturated copper sulphate	+ 0.316	Pipe-to-soil potential measurements

From Table 5.2, the electrode potential for iron in water is approximately – 650mV ref. SCE or – 410mV ref. SHE. On the Pourbaix diagram for iron and water, this point falls within the region of Fe^{2+} (corrosion) telling us that, from a thermodynamic viewpoint, corrosion is likely – which is what we know to be the case! However, starting from this point, let us see what thermodynamic options are open to us to reduce or stop corrosion of iron in water.

a. From the starting point, we see that if the iron electrode potential is made more negative by 'pumping' electrons into it, we arrive in the region of *immunity* or no corrosion. At this potential, the anodic reaction expressed by equation [5.1] is suppressed and cannot take place. The minimum potential depression required for iron in water is approximately –200mV relative to the natural or rest potential. This is called *cathodic protection*.
b. From the starting point, we see that if the iron electrode potential is made more positive by 'sucking' electrons from it, i.e. compelling it to become more anodic, we arrive in the region of *passivity* by forming a tightly adherent and protective layer of Fe_2O_3 (haematite). This process is called *anodic passivation* or *protection* and is a useful and economic method in situations where the medium of immersion is suitable and invariant, e.g. in some chemical and process plant, particularly with stainless steels in sulphuric acid. It is not recommended for sea water however due to the presence of Cl^- which acts as a powerful *depassivator*. Also *potential control* must be precise and accurate otherwise increased corrosion problems can result.
c. From the starting point, we see that by increasing the pH of the water artificially (adding sodium hydroxide for example) the iron can also re-enter the region of *passivity*. This is called *electrochemical* or *pH inhibition* and is a widely

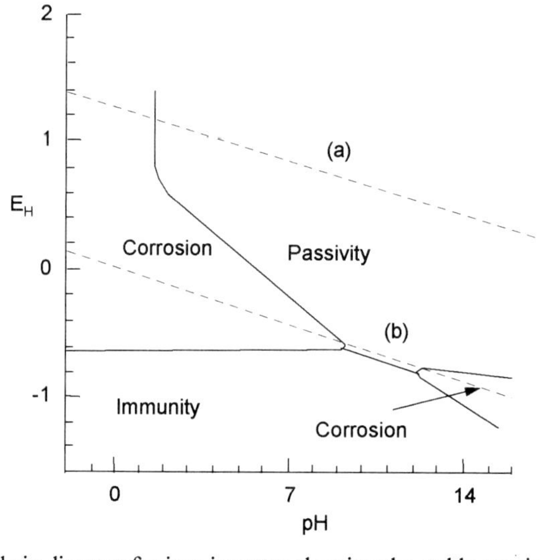

Figure 5.3 Pourbaix diagram for iron in water showing the stable species in each region and the consequences for corrosion reactions.

Table 5.2 Galvanic series of metals and alloys in seawater

Alloy		Potential range on saturated calomel scale (V)
Graphite		+ 0.3 to + 0.2
Platinum	↑	+ 0.35 to + 0.2
Tantalum		About + 0.2
Alloy 6X (passive)	Noble	+ 0.32 to – 0.15
Hastealloy C and C 276		+ 0.10 to – 0.04
Inconel 625		+ 0.10 to – 0.04
Incoloy 825		+ 0.05 to – 0.03
Titanium and titanium alloys		+ 0.06 to – 0.05
22–13–5 Stainless steel (Nitronic 50)		+ 0.08 to – 0.23
EB 26–1		+ 0.17 to – 0.24
Alloy 20 cb3		+ 0.05 to – 0.15
300 Series stainless steel (passive)		–.0.00 to – 0.15
Monel 400 and K–500		– 0.04 to – 0.14
Silver		– 0.09 to – 0.14
17–4 PH stainless steel (passive)		– 0.10 to – 0.20
99.99% Copper		About – 0.14
Inconel 600 (passive)		– 0.13 to – 0.17
Molybdenum		About – 0.17
70–30 Copper-nickel, CDA 715		– 0.13 to – 0.22
Common lead		– 0.19 to – 0.25
Tungsten		About – 0.24
430 and 431 Stainless steel (passive)		– 0.20 to – 0.28
80–20 Copper-nickel, CDA 710		– 0.21 to – 0.27
90–10 Copper-nickel, CDA 706		– 0.21 to – 0.28
Nickel–silver, CDA 752		– 0.23 to – 0.28
Silicon bronze A, CDA 655		– 0.24 to – 0.27
G & M tin bronzes		– 0.24 to – 0.32
Manganese bronze A CDA 675		– 0.25 to – 0.33
410 Stainless steel (passive)		– 0.24 to – 0.35
Inhibited admiralty, CDA 443, 444 and 445		– 0.25 to – 0.34
Lead–tin solder 50/50		– 0.26 to – 0.35
ETP copper, CDA 110		– 0.28 to – 0.36
Red brass, CDA 230		– 0.20 to – 0.40
17–4 PH stainless steel and alloy 6X (active)		– 0.20 to – 0.40
Cast brasses and bronzes		– 0.24 to – 0.40
Naval brass, CDA 464		– 0.30 to – 0.40
Aluminium bronze D, CDA 614		– 0.30 to – 0.40
Inconel 600 (active)		– 0.30 to – 0.42
Austenitic nickel cast iron		– 0.35 to – 0.47
300 series stainless steel		– 0.35 to – 0.57
410, 430 and 431 stainless steels (active)		– 0.45 to – 0.57
Maraging steels		– 0.57 to – 0.58
4130 alloy steels		About – 0.6
HY 80 and high strength steels		– 0.60 to – 0.63
Low alloy steels		– 0.57 to – 0.63
Plain carbon steels	Active	– 0.60 to – 0.70
Cast irons		– 0.60 to – 0.72
Aluminium alloys	↓	– 0.70 to – 0.90
Zinc		– 0.98 to – 1.03
99.99% Aluminium		– 1.25 to – 1.50
Magnesium		– 1.60 to – 1.63

used method of iron corrosion control in (i) soils around oil and gas pipelines – by adding lime backfill, and (ii) cooling systems of engines, generator and pumping sets, etc.

However, *Pourbaix* diagrams do not give any indication of corrosion rates and are generally only widely available for pure water/metal systems – not sea water. To get some indication of corrosion rates we use polarisation (*Evans*) diagrams.

5.4.2 Polarisation (Evans) diagrams

To understand the use of these diagrams let us firstly visualise the situation of an anodic and cathodic area on the surface of a metal. If we prevent electronic flow by insertion of a massive resistance *R*, the potentials of the anodic and cathodic surfaces are x and y respectively in Fig. 5.4.

However as this artificial resistance is reduced, current starts to flow between sites (anode to cathode) until dynamic equilibrium is reached at E_m – the *mixed corrosion potential* (*electrode potential*). The corrosion rate of the metal at this point is reflected by the current flow i_m. Line x-m represents *anodic dissolution* and line y-m represents *cathodic oxygen reduction reaction*. In reality this situation of perfect dynamic equilibrium at the E_m end point is quite rare due to the metal internal resistance, higher resistance media, lack of oxygen supply etc. Hence corrosion currents and rates are slightly lower than theoretically predicted. When cathodic protection is applied, line y-m extends to point n due to promotion of the cathodic reaction. At point n the anodic current reduces to zero and in represents the applied cathodic current.

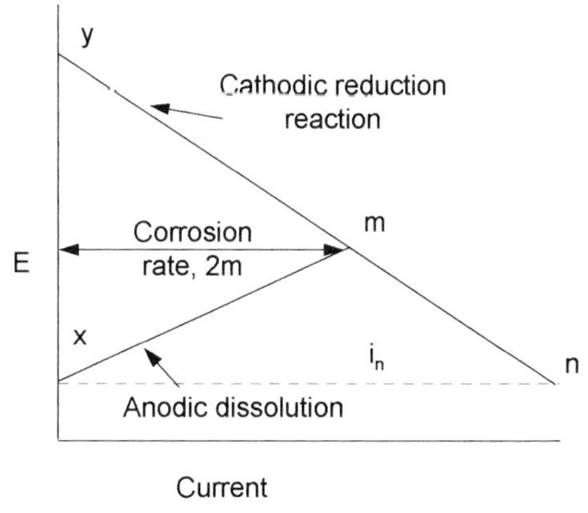

Figure 5.4 Polarisation/current diagram for a metal surface with no resistance polarisation effects at mixed potential.

5.5 Conclusions

Any overview of aqueous corrosion must reflect the known evidence that throughout the economy as a whole, the wider application of known knowledge and techniques of corrosion control would save approximately the costs referred to in Section 5.1. It is a depressing reflection of modern society that corrosion costs as a proportion of the GDP seem to have little changed in the past 25 years or so. The subject of corrosion still seems to remain a '*cinderella*' – but costly '*cinderella*' – subject when compared with other engineering considerations. Furthermore it is not particularly surprising to conclude that amongst the vast majority of corrosion mechanisms, aqueous or some other related process are important, i.e. *pitting* and *crevice-type* mechanisms. Often there is a perception of some general aqueous corrosion issue which has to be tackled through corrosion control measures, when in fact what is being faced in reality is mobile localised corrosion. This *mobility* may be variable in rate with time and consequently selection of corrosion control techniques has to be undertaken with considerable care to increase probability of success throughout the intended operating life of the particular system being considered.

References

1. Jarman S and Burstein (Eds), Corrosion, **1** and **2**, Third edition, Butterworth Heinman.
2. Fontana M G, Corrosion Engineering, McGraw Hill, Singapore, 1987.

Chapter 6

Protection by organic coating against corrosion: an overview

W Funke – University of Stuttgart

6.1 Introduction

Developments in organic coatings for corrosion protection of metals during the last decade are characterized by:-

- Environmental requirements and legislation to protect against health hazards in the production, application and repair of corrosion protective paints or coatings.
- New insights into the protective mechanisms and the requirements for optimal protection.

6.2 History and problems

Classical protective coating systems have been *linseed oil/red lead* and *coal tar*. Both represent the two important protective mechanisms: *electrochemical* and *barrier*. As the drying time of linseed oil, which forms films by a complicated oxidative process, was rather long, industrial mass production demanded new binders with faster film formation. Accordingly natural oils were substituted by alkyd resins, vinyl polymers and finally epoxy resins and polyurethanes.

In addition to red lead, *zinc chromates* were used as anticorrosive pigments. However, in paint science and technology the classical protective concept, represented by linseed oil/red lead, was still largely maintained, even though, often, the requirements enabling one protective mechanism to operate might prevent or hinder the other mechanism. Positive results claimed for such coating systems are not surprising, because it is possible that one mechanism may work efficiently at the expense of the other to yield a satisfactory result. A typical example is the combination of a polyurethane binder, which has a low permeability to water, and zinc chromate. Despite this pigment belonging to the most efficient anticorrosive pigments, the binder may encapsulate it so closely that its soluble fraction cannot diffuse to the metal substrate. It is imperative therefore to know the principal requirements for a protective mechanism to optimize its efficiency and avoid useless paint components.

6.3 Requirements for protective coatings

For *electrochemical* protection by anticorrosive pigments or by corrosion inhibitors to be used in organic coatings, the binder should be permeable to water and to the soluble active fraction of the anticorrosive agent.

Despite much work on the permeability of organic coatings for water, oxygen and even ions from the environment, little is known on permeability requirements for chromate, phosphate and other effective ions of anticorrosive pigments. It is well known, however, that pigment volume concentration is closely related to the protective efficiency of such pigments. This relationship indicates strongly the importance of the permeability of the coating film for active ions.

As a consequence anticorrosive pigments should only be used with binders, which are sufficiently swellable in water and permeable to the soluble fraction of the pigment. Protective coatings of this type, of course, cannot be expected to exhibit good wet adhesion. However, it is important that adhesion can be regenerated when the coating dries again and that this regeneration is repeatable to some extent. As regeneration of adhesion normally decreases with the exposure events, simple protective coatings, based on the electrochemical mechanism, are not recommended for heavy duty applications.

This statement is emphasized by the fact that corrosion-stimulating anions from the environment, such as chloride and sulphate, strongly inhibit the protective action of the anticorrosive pigment [1].

On the other hand anticorrosive pigments or corrosion inhibitors are indispensable for protective coating systems based on aqueous polymer solutions or dispersions, which dry at ambient temperatures.

Organic coatings which operate through the *barrier mechanism*, rather than by electrochemical protection, should have a very low permeability to water, oxygen and ions and adhere to the metal support even on exposure to high humidity or water. The latter requirement, *wet adhesion*, is not easy to achieve; adhesion bonds are at metal surfaces due to the natural oxide layer, are strongly polar and the bonding sites or chemical groups are sensitive to water molecules penetrating to this interface. How can this problem be solved?

6.4 Cooperation of adhesive bonds

The usual way is to use *binders*, which by their very chemical structure have little affinity for water or which are cross-linked densely enough to decrease or prevent permeation. The first approach is limited, because usually the same binder molecules have to provide adhesion, i.e. must be polar, and this means sensitivity to water.

Cross-linking is a more efficient means to decrease permeability, however, caution is required because cross-linking strongly influences mechanical properties of the coating. Permeability is efficiently decreased by prolonging the diffusion pathways through the coating film by using "barrier pigments", such as aluminium flakes or talc [2]. However, a disadvantage is that these pigments also reduce the rate of film formation by retarding the evaporation of solvents or water. The chosen route could be solventless systems, if viscosity requirements can be obeyed. It is also important that during film formation the particles of the barrier pigment should align the flat part of their surface parallel to the metal substrate. Otherwise such pigments may increase permeability substantially.

Finally it should be noted that to increase the diffusion pathway effectively by barrier pigments, the pigment/binder interaction should be resistant to water molecules, much the same as is required for the coating/film substrate interface.

This requirement leads us back to the interface at the coating/metal surface and to the problem of how to achieve good wet adhesion when water sensitive, polar adhesion bonds to the substrate are principally needed for the adhesion. To solve this problem the structure and conformation of the binder molecules, which happen to adjoin the metal surfaces, have to be considered. As polar groups are needed for adhesion the question is how to stabilize this interaction against interference by water molecules?

A logical premise for corrosion protection by coatings is that no corrosion can occur as long as the bonds of the coating film to the metal surface remain intact. As there are good reasons to assume that covalent bonding plays only a minor part in adhesion to metals such as steel, where in fact mainly *secondary valence forces*, hydrogen bonds and ionic bondings provide adhesion. These bonds, though

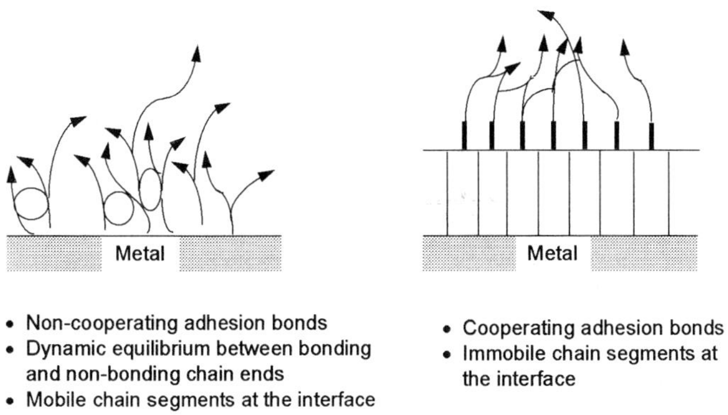

Metal
Metal
• Non-cooperating adhesion bonds
• Dynamic equilibrium between bonding and non-bonding chain ends
• Mobile chain segments at the interface
• Cooperating adhesion bonds
• Immobile chain segments at the interface

Figure 6.1 Non-cooperative and cooperative adhesion bonds at the coating/support interface.

dynamic by nature, may be stabilized by their attachment to rigid backbone chains of the binder. This forces them to cooperate (Fig. 6.1). A bond may only disrupt if neighbouring bonds at the same polymer chain are simultaneously disrupted. For this purpose, the macromolecular chains adjoining the substrate should be immobilised and rigid and this is possible by cross-linking. An indication of chain mobility is the *glass transition temperature* of the cured binder, T_g. T_g should be substantially above the performance temperature range of the coating, and this also in the presence of water, i.e. in the "swollen state" of the coating film. It can be shown that wet adhesion improves significantly with an increase of T_g [3,4] (Table 6.1). Rigidity of the backbone chain is related to film formation, i.e. cross-linking. If the binder molecules are already rigid before coming in contact with the substrate, the adsorptional adjustment to the surface morphology may be difficult and therefore the degree of interaction lowered.

Rigidity of coating films necessarily affects their mechanical properties, e.g. brittleness may be a consequence. The thicker the film is, the more this problem is aggravated. This situation forces consideration of the relative importance of various coating properties and, especially, whether base coats of usual thickness are needed for good adhesion. As there is no doubt that adhesion to the support is the most important coating property, why not provide optimum conditions for this property?

As adhesion is an interfacial property and the thickness of the organic layer directly involved is much less than that of the usual base coats, film thickness should be decreased to a level where mechanical stresses even in rigid films can be resisted. It is well known that comparatively brittle coatings may resist mechanical forces when their thickness is very low. An additional advantage of thin adhesion layers is that film formation is faster and curing more complete as compared with thick films. It can be shown that wet adhesion of coatings with normal (ca. 40 μm) thickness applied to suitable very thin (ca. 0.2 μm) organic adhesion layers exhibit remarkably higher wet adhesion [4,5] (Fig. 6.2).

Table 6.1 Wet adhesion and glass transition temperature of organic coatings in dry conditions

Coating No	Wet adhesion (hrs)	T_g (°C)
I	1,880	88
II	1,880	83
III	1,008	82
IV	120	30/83
V	18	44
VI	8	82
VII	1.5	20
VIII	1.0	27

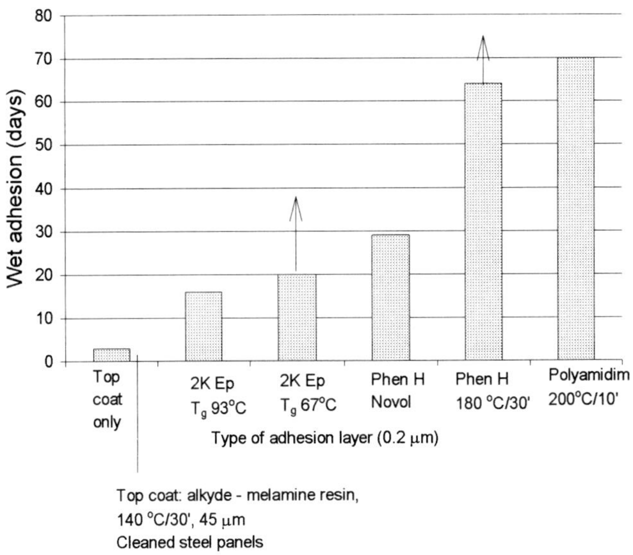

Figure 6.2 Influence of several thin adhesion layers (0.2 μm) on wet adhesion of an alkyd/melamine resin film.

6.5 Conformation of macromolecules at metal surfaces

As normal organic base coats have a film thickness between 20–40 μm, the respective paints are rather concentrated solutions. After their application to a substrate, binder molecules in the neighbourhood of the substrate inevitably compete for adsorption sites at the surface and the number of adhesion bonds per macromolecule is small (Fig. 6.3). It is reasonable to assume that adhesion improves as more polar groups of a macromolecule interact with the surface. For more effective adsorption the paint should be much more dilute and pigments are superfluous in the adhesion layer. Moreover, it is questionable whether polar groups to this extent are also needed within the bulk film. They will interact there with each other, e.g. by hydrogen bonds, thus contributing to cross-linking to some degree. However, likewise, they render the film hydrophilic and disruptive in the presence of water. It would be better to adsorb hydrophilic macromolecules directly on the substrate surface and use less hydrophilic species for the subsequent coating layers.

It was quite surprising that applying macromolecules of the polyacrylic acid type from very dilute aqueous solutions to steel surfaces, increases wet adhesion of successive different coating layers of normal film thickness quite substantially with a maximum of efficiency at concentration yielding adhesion layers of a thickness as low as 10–20 nm (Fig. 6.4). In these dilute solutions binder molecules have enough adsorption area available to decrease competition with neighbouring

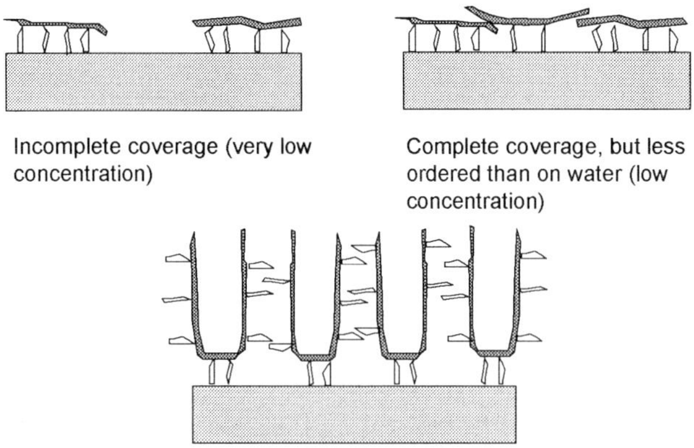

Figure 6.3 Concentration effects of polar macromolecules at polar substrates.

molecules at the metal surface to a minimum, but still cover the substrate sufficiently. Of course polyacrylic acid is not an ideal binder for adhesion layers. A disadvantage is that the subsequent coating layers of normal thickness must be solvent-borne and not water soluble. This situation is a challenge to synthesize binders for ultra thin adhesion layers, which may be cross-linked to fix them to the substrate by cooperative bonding.

This Section on corrosion protection by organic coatings may be summarized as follows:

i. Cooperative bonding improves wet adhesion and delays delamination by water, a process preceding corrosion.
ii. Suitable conformational arrangements of macromolecules in thin adhesion layers increases the bonding efficiency and thereby wet adhesion.
iii. Thin adhesion layers of specially synthesized, cross-linkable macromolecules

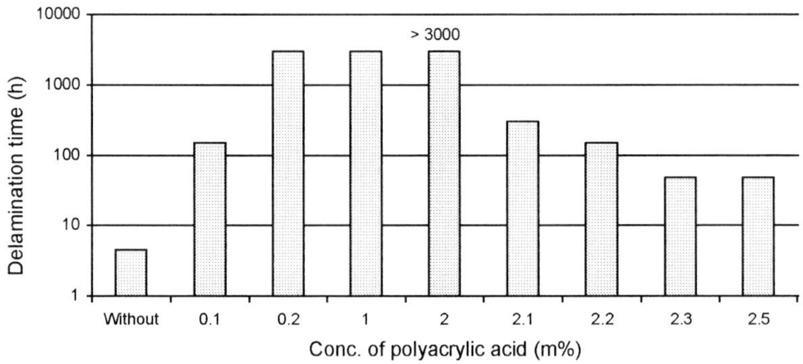

Figure 6.4 Wet adhesion and concentration of polyacrylic acid (PAA, 23°C) – alkyd/melamine resin top coat.

combined with other requirements for the barrier-type of protection may become an attractive and promising alternative to *electrochemical-type*, especially for heavy duty applications.

6.6 Practical consequences

In addition some practical recommendations are noted:

- It is *not* reasonable to attempt to combine the electrochemical and barrier mechanisms in the same base layer on the metal substrate.
- It is possible to apply a coating containing an anticorrosive pigment on a pigment-free adhesion layer to provide protection at paint defects and damage.
- Coating systems for cathodic protection by external electrical potentials should be as impermeable as possible, i.e. the barrier-type of coating system is preferred.
- Zinc-rich base coats have to be protected against aggressive agents from the environment by barrier top coats. Zinc-rich base coats should only become active and protect at sites where coating systems have been damaged.

Acknowledgement

The support of the Arbeitsgemeinschaft fur industrielle Forschungsvereingung e.V. and the Bundesministerium fuer Wirtschaft is gratefully acknowledged.

References

1. Reichle P and Funke W, *farbe + lack*, **93**, p537, 1987.
2. Funke W, *farbe + lack*, **89**, p86, 1983.
3. Funke W, *farbe + lack*, **93**, p721, 1987.
4. Funke W, *The Electrochem. Soc. Proc.*, **89**, 13, p121.
5. Neidhammer H and Oliveira L, *farbe + lack*, **96**, p99, 1990.

Chapter 7

High temperature gaseous and molten salt corrosion

K Natesan – Argonne National Laboratory, USA

7.1 Introduction

Corrosion of materials at elevated temperatures is a potential problem in many systems within the chemical, petroleum, process and power generating industries. The corrosion phenomenon involves interaction between the structural material and the exposure environment. The interactions are generally undesired chemical reactions that can lead to wastage and alter the structural integrity of the materials. Therefore, material selection for high temperature applications is based not only on a material's strength properties but also on its resistance to the complex environments prevalent in the anticipated exposure ambient. As plants become larger, satisfactory performance and reliability of components play a greater rôle in plant availability and economics. However, design personnel are becoming increasingly concerned with finding the least expensive material that will satisfactorily perform the design function for the desired service life. High temperature corrosion can be broadly classified into two distinct environments: (a) *gaseous corrosion in atmospheres that include single and multiple oxidants*; and (b) *deposit-induced corrosion in environments that include solid and liquid deposits* which comprise ash/slag constituents, solid sorbents and molten salts of various types.

7.2 Corrosion in single oxidant environments

Considerable research on the causes, effects, and prevention of different types of corrosion has been under way for many years. However, when one studies reaction kinetics that involve only one reaction equilibrium, such as H_2O/H_2 or CO_2/CO for oxidation, H_2S/H_2 for sulphidation, CH_4/H_2 for carburization, or NH_3/H_2 for nitridation, the reaction potential of the participating species can be uniquely established by the standard free energy of formation for the reaction. For an oxidation reaction of a pure metal M, such as

$$M + \frac{1}{2} O_2 = MO \qquad [7.1]$$

the oxygen partial pressure (pO_2) for the M/MO equilibrium given by

$$pO_2 \text{ at } M/MO \text{ phase boundary} = e^{\frac{\Delta G^o}{RT}} \qquad [7.2]$$

where ΔG^o is the free energy change for the M/MO equilibrium temperature T and R is the universal gas constant. In binary and ternary alloys, the activities of the reactive element (a_M) and the oxide (a_{MO}) should also be considered. In such cases, *Eq. 7.2* becomes

$$pO_2 \text{ at } M/MO \text{ phase boundary} = \frac{a_{MO}}{a_M} e^{\frac{\Delta G^o}{RT}} \qquad [7.3]$$

In general, element activity in an alloy is given by

$$a_M = \gamma_M X_M \qquad [7.4]$$

where γ_M and X_M are the *activity coefficient* and *mole fraction* of M in the alloy, respectively. If a value for the pertinent γ_M is not available, ideal behaviour is assumed and γ_M is assigned a value of unity.

Under similar single oxidant conditions, the oxidation, sulphidation, carburization, nitridation or halogenation of a metal can be established by the standard free energy formation of the metal oxide, sulphide, carbide, nitride or halide. Plots of standard free energy of formation versus temperature for compounds such as oxides, sulphides, carbides, nitrides or halides have been developed in the literature and serve as a guide in establishing the relative stability of the corrosion product layers with respect to both the partial pressure of the reactive species in the exposure environment and the alloy composition. Figures 7.1–7.3 show the thermodynamic stability of several oxide, carbide and nitride phases that form by reaction between oxygen, carbon or nitrogen and the constituent of various structural materials such as heat resistant alloys, superalloys, and refractory metals. As a result, in classical studies on materials exposed to single oxidants, the central theme is to understand the *kinetic* aspects of continuous scale development and the *mechanisms of scale growth* as a function of variables such as temperature, time and reactant activity in the exposure environment. An enormous amount of literature exists on nucleation and growth of corrosion product scales for a wide range of materials exposed to several single oxidant environments.

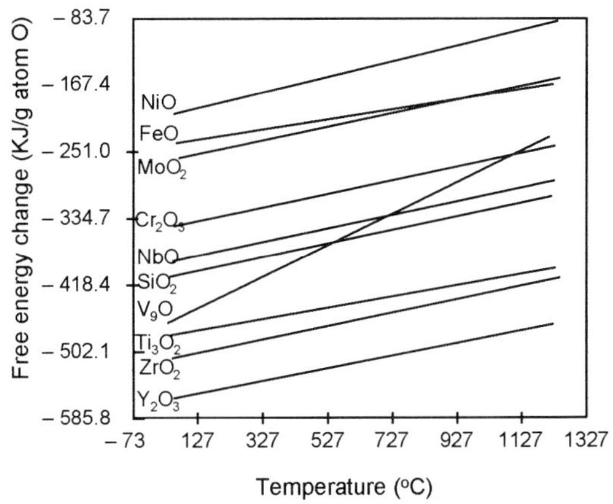

Figure 7.1 Standard free energy of formation of metal oxides as a function of temperature.

In addition, the effort has concentrated on modifying alloy compositions to: (a) form slower growing scales (for example, alumina, chromia, or silica scales instead of iron oxides, NiO, or CoO in heat resistant iron-, nickel-, and cobalt-base alloys); (b) reduce corrosion rates (smaller parabolic rate constants); (c) improve adhesion of scales to substrates (by additions of reactive elements such as Y, Ce, and La), especially under thermal cycling conditions; and (d) provide surface enrichment of structural alloys with elements (such as Cr, Al, and Si on low and intermediate chromium steels and austenitic alloys) that form slow growing scales.

Even though corrosion of materials in single oxidant environments is fairly simple to examine and provides a basis for formulation of alloys solely on the basis of their oxidation resistance it is very limited in practical systems in view of the complex nature of the environments in these systems.

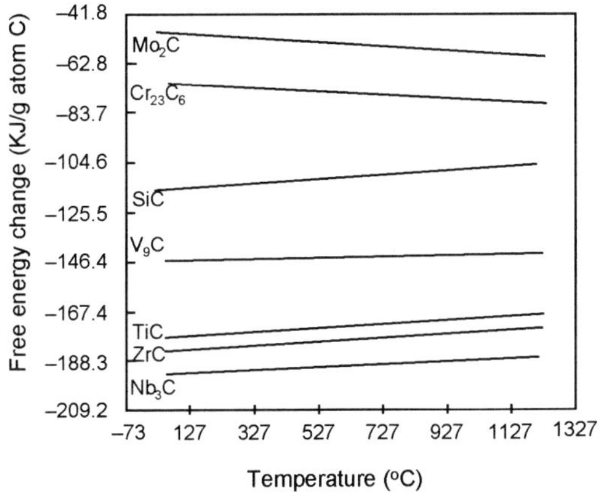

Figure 7.2 Standard free energy of formation of metal carbides as a function of temperature.

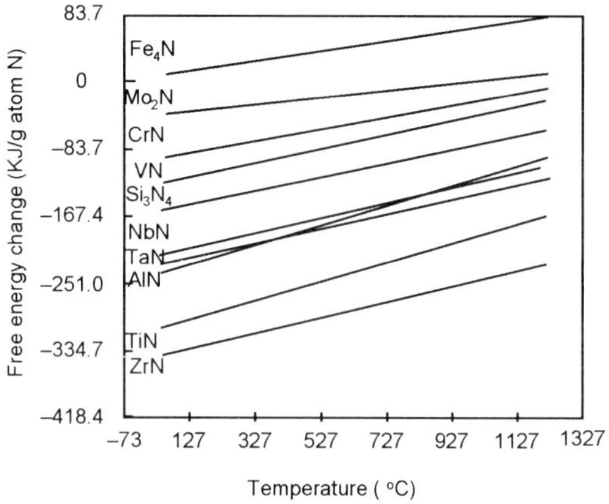

Figure 7.3 Standard free energy of formation of metal nitrides as a function of temperature.

7.3 Thermodynamic aspects of reaction in mixed oxidants

When materials are exposed to multicomponent gas environments or mixed oxidants, several of the fundamental processes mentioned above can occur simultaneously. Furthermore, the second reactant may either modify or degrade the corrosion product layers that form by reaction of the first reactant and the substrate elements. In these situations, reactions will occur not only at the gas/scale interface, but also within the scales that develop on the surface and in the base metal beneath the reaction products. As a result, the development of suitable and meaningful thermodynamic and kinetic frameworks for both nucleation and growth of scale layers on materials exposed to mixed oxidant atmospheres is quite complex.

When studying reactions between metallic alloys and complex gas environments, the chemical activities of several reactive species in the gas phase must be considered simultaneously, and these activities are generally established by gas phase equilibria, especially at elevated temperatures. In binary and ternary gas mixtures, the chemical potentials of the reactive species can be readily established, as a function of temperature, from the room temperature gas composition and free energy data for reactions. For a gas system that consists of CO, CO_2, H_2, H_2O, CH_4, H_2S, and N_2 or NH_3, several gas equilibria have been used to formulate a set of non-linear algebraic equations [1,2]. One can use iterative procedures to determine gas compositions at elevated temperatures that yield minimum free energy for the system and satisfy the conservation of different reactive elements (i.e. C, H, S, O and N) and the total pressure of the gas mixture. Results of such analyses have been used to establish the elevated temperature gas compositions, the partial pressures of oxygen, sulphur and nitrogen, and carbon activity in the gas mixture. In the analysis of gas/metal interactions in mixed gas atmospheres, it is inherently assumed that equilibria among various molecular gas species prevail and chemical activities are calculated by using the above approach or some similar method.

Pettit et al [3] provided an excellent review of the thermodynamic aspects of metal corrosion in a bioxidant situation. In an oxygen/sulphur environment, on the basis of their approach, the following surface reactions are possible on a divalent metal *A*:

$$A + \frac{1}{2}O_2 = AO \qquad [7.5]$$

$$A + \frac{1}{2}S_2 = AS \qquad [7.6]$$

where *AO* and *AS* are the *oxide and sulphide reaction products*, respectively. Under equilibrium conditions, the *activities* and *partial pressures* of oxygen and sulphur are defined by the relations

$$(a_O)_{eq} = pO_2^{1/2} = e^{\frac{\Delta G_{AO}}{RT}} \qquad [7.7]$$

$$(a_S)_{eq} = pS_2^{1/2} = e^{\frac{\Delta G_{AS}}{RT}} \qquad [7.8]$$

where ΔG_{AO} and ΔG_{AS} are the *standard free energies of formation of the oxide and sulphide*, respectively, at temperature T. From *Eqs. 7.7 and 7.8*, one should be able to deduce the conditions for oxidation or sulphidation; however, a further reaction must be considered, namely

$$AS + \frac{1}{2}O_2 = AO + \frac{1}{2}S_2, \qquad [7.9]$$

with the equilibrium condition

$$\left(\frac{a_S}{a_O}\right)_{eq} = \frac{a_{AS}}{a_{AO}} e^{\frac{\Delta G_{AS} - \Delta G_{AO}}{RT}} \qquad [7.10]$$

If we assume unit activity for the phases AS and AO, *Eq. 7.10* can be reduced to

$$\left(\frac{a_S}{a_O}\right)_{eq} = e^{\frac{\Delta G_{AS} - \Delta G_{AO}}{RT}} \qquad [7.11]$$

Examination of *Eqs. 7.7, 7.8 and 7.11* permits the identification of various situations that limit the type of surface corrosion products that can be formed, as follows:

i. If $(a_O)_{gas} > (a_O)_{eq}$ and $(a_S)_{gas} < (a_S)_{eq}$, then *AO is the only stable surface phase*;

ii. If $(a_O)_{gas} < (a_O)_{eq}$ and $(a_S)_{gas} > (a_S)_{eq}$, then *AS is the only stable surface phase*; and

iii. If $(a_O)_{gas} > (a_O)_{eq}$ and $(a_S)_{gas} > (a_S)_{eq}$, then *both* AO and AS should be stable, and form as surface products. However, reference to *Eq. 7.11* indicates that *only one phase will form*, depending on which of the following conditions prevail:

a. $$\left(\frac{a_S}{a_O}\right)_{gas} > \left(\frac{a_S}{a_O}\right)_{eq}$$

This condition will cause *reaction 7.9* to proceed to the left, and *AS will be the stable phase* where the metal is in contact with the gas phase.

b. $$\left(\frac{a_S}{a_O}\right)_{gas} < \left(\frac{a_S}{a_O}\right)_{eq}$$

In this case, *AO will be the stable phase*, and *reaction 7.9* will proceed to the right. A similar analysis can be made to evaluate the effects of other environments (carbon/oxygen, carbon/sulphur, etc.) on scale formation.

A convenient way of representing the possible corrosion products as a function of gas chemistry is to construct thermochemical diagrams that depict the stability ranges of various condensed phases as functions of the thermodynamic activities of the two components of the reactive gas. A schematic of the thermochemical diagram for Fe, Cr, Ni and Mn and their oxides and sulphides at 875°C is shown in Fig. 7.4 as a function of oxygen and sulphur partial pressures. The diagram shows that at oxygen and sulphur pressures below the metal/oxide and metal/sulphide boundaries, respectively (*region A* of upper diagram), the *metal will be stable* and undergo neither oxidation nor sulphidation reaction. In *regions B and C*, the oxygen or sulphur partial pressures are sufficient to form *oxide* and *sulphide phases*, respectively. In *region D*, oxide is the thermodynamically stable phase; however, the sulphur partial pressure is sufficient to form *sulphides* in the subscale. The reverse is true in *region E*.

In the construction of these diagrams, the thermodynamic activities of the metal and corrosion product phases are assigned a value of unity. Figure 7.5 shows a thermochemical stability diagram for the chromium–carbon–oxygen system developed for a temperature of 982°C. In multicomponent alloys, the activities of

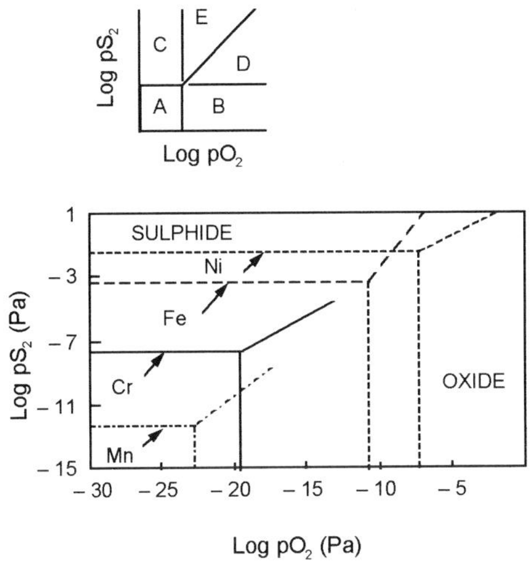

Figure 7.4 Thermochemical diagrams for M–S–O systems at 875°C (where M = Cr, Fe, Ni and Mn).

constituent elements will be less than unity and should be accounted for in the analysis. Further, in gas/solid reaction that involve multicomponent alloys, a more complex corrosion product (i.e. more complex than a binary compound) can result and will also decrease the thermodynamic activities of the specific phase in the mixture. This is especially true in cases where the corrosion reactions lead to a liquid phase.

Natesan [4,5] used this thermodynamic approach to determine the extent to which fundamental processes can influence materials in selection in *coal-conversion applications*. Figure 7.6 shows oxygen/sulphur thermochemical diagrams that depict the stability of different phases in Type 310 stainless steel (an alloy of composition Fe25Cr20Ni) as a function of oxygen and sulphur pressures at 727, 927 and 1127°C. Also shown in this figure are the oxygen and sulphur partial pressures calculated from the estimated gas compositions in various high-Btu *coal gasification* processes operating with low sulphur coal feedstock. The diagram clearly shows that the gas environments in different processes are such that $(a_O)_{gas} < (a_O)_{eq}$, $(a_S)_{gas} > (a_S)_{eq}$, and , $(a_S/a_O)_{gas} < (a_S/a_O)_{eq}$ (*condition iiib*); therefore, based on thermodynamic considerations, high Cr alloys should develop protective Cr-rich oxide scales with sulphides of Cr and Fe beneath the scale. Another important use of these diagrams is in *a priori* prediction of liquid corrosion product phases. For example, Fig. 7.7 shows thermochemical diagrams for Inconel 671 (an alloy of composition Ni48Cr) at three different temperatures as a function of oxygen and sulphur partial pressures. The figure shows that the high Cr content promotes the formation of Cr oxide scale at 1127°C; however, the formation of Ni/Ni sulphide eutectic with a melting temperature of 645°C is a distinct possibility at lower temperatures.

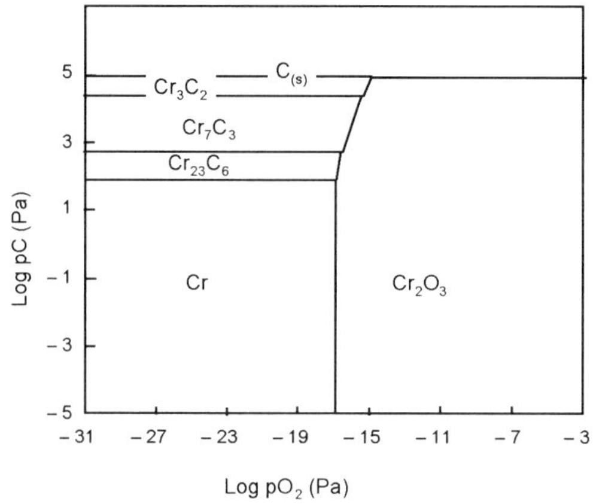

Figure 7.5 Thermochemical diagram for Cr–C–O systems at 982°C.

7.4 Scaling of alloys in bioxidant atmospheres

The formation and maintenance of protective surface oxide scales are essential in preventing rapid degradation of structural alloys used in elevated temperature applications. In principle, one of three oxides (Cr_2O_3, Al_2O_3 or SiO_2) constitutes the major component of a thermally formed surface scale. In order to understand the development of protective oxide scales on a given alloy, it is essential to establish the influence of pO_2 variation in the gas environment on the morphology of corrosion product layers. The interrelationships among partial pressures of oxygen and of the second reactant such as sulphur, carbon, nitrogen or halide in the gas phase, the alloy chemistry, and the test temperature and pressure are of considerable interest in establishing the critical oxygen pressures (if any) that are required to achieve a stable, adherent oxide scale on the metal surface.

7.4.1 Behaviour of Cr_2O_3-forming alloys

Cr_2O_3 is the predominant constituent of surface scales formed on heat resistant austenitic Fe–Cr–Ni alloys. Under isothermal conditions, alloys containing a minimum of about 14 wt.% Cr can be expected to form protective Cr-rich oxide scales in oxygen-rich (single oxidant) environments at temperatures of 627–827°C, while about 20 wt.% Cr is required at temperatures of about 1027°C. In oxygen/sulphur mixed gas environments, typical of those encountered in coal gasification and combustion atmospheres, experience shows that a thermodynamically stable protective oxide scale may not form because of the sulphur in the gas phase, even though the thermodynamic stability of the oxide and sulphide phases would dictate oxide formation (see Fig. 7.8 for data shown for Type 310 stainless steel). Figure 7.9 shows the corrosion scale morphologies developed on Type 310 stainless

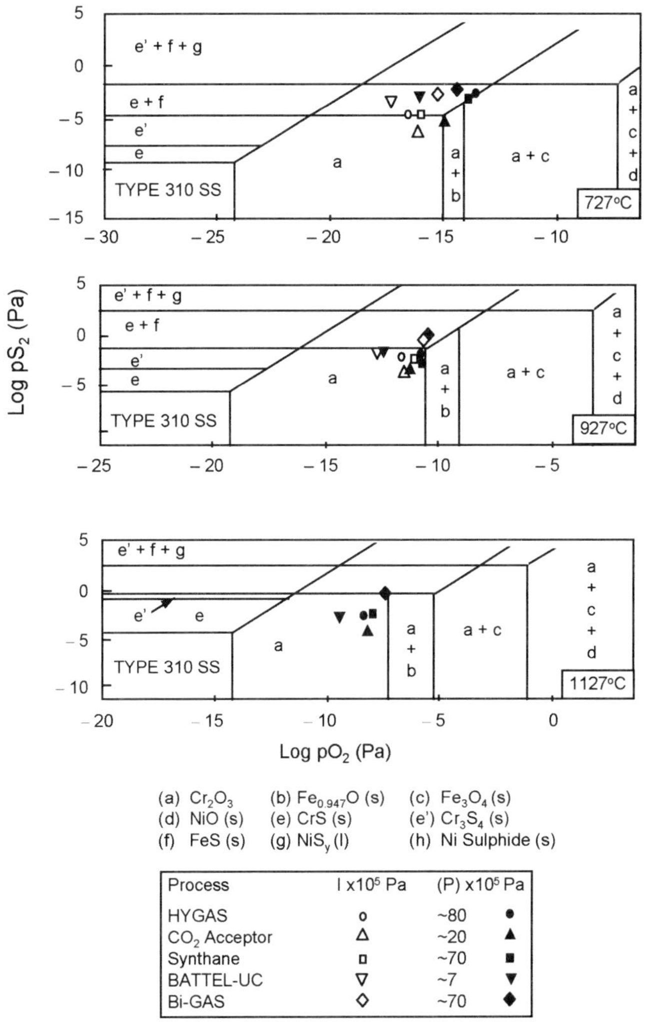

Figure 7.6 Oxygen/sulphur thermochemical diagram for type 310 stainless steel at 727, 927 and 1127 °C, showing gas environments calculated for several coal gasification processes.

steel specimens after exposure at 875°C for 168 h in four different complex gas environments with nominally the same pS_2 but with different pO_2 values.

In practice, an *excess* of oxygen above the level defining thermodynamic equilibrium between Cr_2O_3 and "CrS" is required to form Cr_2O_3 as a continuous surface layer. For example, the corrosion behaviour of Alloy 800 (with a composition of Fe21Cr32.5Ni) exposed to oxygen–sulphur atmospheres has been depicted as a function of *excess oxygen parameter*, $pO_2/pO_2(eq)$, where pO_2 and $pO_2(eq)$ are the oxygen partial pressures in the gas mixture and that corresponding

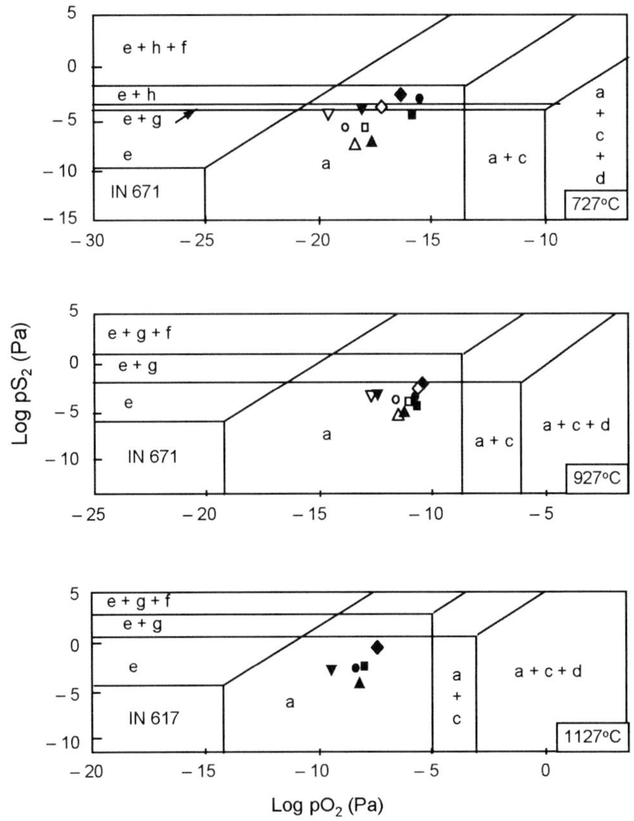

(a) Cr_2O_3 (b) $Fe_{0.947}O$ (s) (c) Fe_3O_4 (s)
(d) NiO (s) (e) CrS (s) (f) FeS (s)
(g) NiS_y (l) (h) Ni Sulphide (s)

Process	I x10^5 Pa		(P) x10^5 Pa	
Hygas	o		~80	●
CO_2 Acceptor	△		~20	▲
Synthane	□		~70	■
BATTEL-UC	▽		~7	▼
Bi-Gas	◇		~70	◆

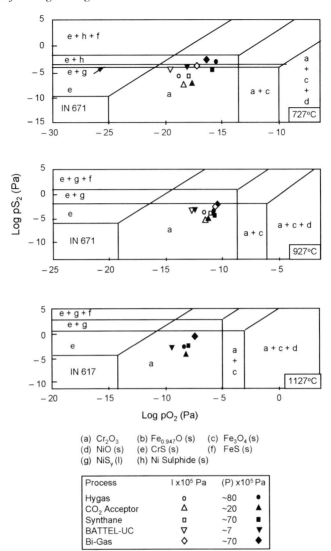

Figure 7.7 Oxygen/sulphur thermochemical diagrams for IN 671 alloy at 727, 927 and 1127°C, showing gas environments calculated for several coal gasification processes.

to chromium oxide/chromium sulphide equilibrium, respectively. Figure 7.10 shows the effect of variation in pO_2/pO_2(eq) ratio on the scale of thickness and penetration depth for the alloy after a 25 h exposure, at temperatures between 750 and 1000°C, to gas mixtures with a wide range of oxygen and sulphur partial pressures. Results from these tests also indicate the "*transition*" or "*kinetic*" boundary for oxide formation in the high chromium alloys is at a threshold pO_2 » 10^3 times the pO_2 values for Cr oxide/Cr sulphide equilibrium.

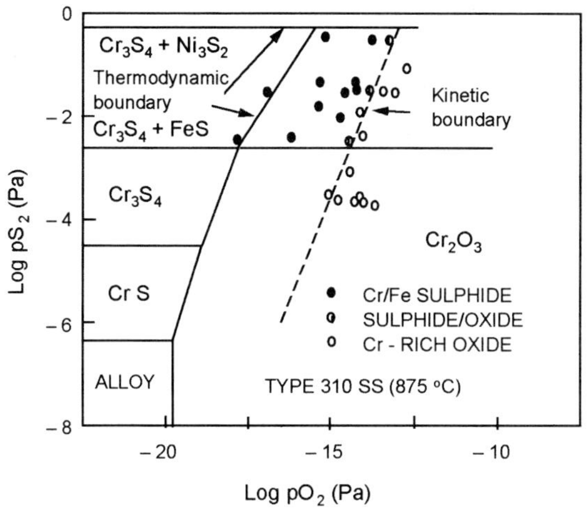

Figure 7.8 Type of scale developed on Type 310 stainless steel as a function of oxygen and sulphur partial pressures in gas environment at 875°C.

LaBranche et al [6] and Natesan [7] observed a transition from non-protective oxidation of pure chromium at oxygen partial pressures 30 and 42 times that for Cr oxide/Cr sulphide equilibrium. Such a large deviation in pO_2 *cannot* be explained in terms of solid solution effects or reduction in activity of sulphide to values below unity. Furthermore, the effect of lowering the exposure temperature is to increase the threshold pO_2 for oxide formation. For example, at 650°C the absolute values of pO_2 for protective oxide formation are approximately five orders of magnitude higher than the thermodynamic equilibrium value. The consequences of oxygen pressures lower than the threshold value in the environment are increased rates of scale growth and substantially deeper penetration of sulphur into the alloy substrate.

In contrast with the corrosion processes that lead to predominantly solid and sometimes liquid (at elevated temperatures) corrosion product phases upon exposure of metallic alloys to oxygen/sulphur atmospheres, exposure of alloys to environments such as oxygen/chlorine and oxygen/hydrogen chloride mixed gases can result in *volatile* corrosion product phases and thereby accentuate corrosion wastage. For example, McNallan et al [8] and Ihara et al [9] reported formation of volatile metal chlorides at the scale metal interface, as well as vapour phase transport of these chlorides through the porous oxide layers, as the major cause of accelerated corrosion rates observed in chlorine-containing mixed gas atmospheres.

7.4.2 Régimes of corrosion behaviour for high chromium alloys

On the basis of the morphological information developed on a number of commercial engineering alloys, advanced highly alloyed metallic materials, and

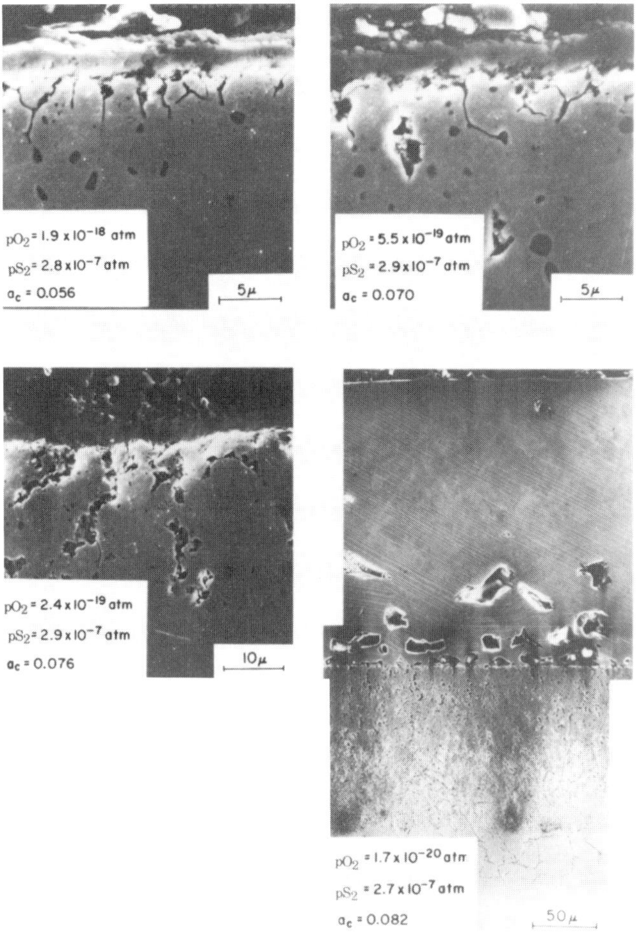

Figure 7.9 Scanning electron micrographs of cross-sections of Type 310 stainless steel specimens exposed at 875°C for 168 h in several oxygen/sulphur mixed gas environments.

model alloys exposed to oxygen/sulphur mixed gas environments, three régimes can be defined (see Fig. 7.11) to describe the oxidation/sulphidation behaviour of the alloys at elevated temperatures [10]. In *régime 1* below the base-metal sulphidation potential and the threshold oxide formation boundary, high chromium alloys develop outer scales of chromium sulphide with variable amounts of soluble Fe and Co. The substrates contained substantial porosity and internally sulphidized chromium. *Régime 2* conditions, above the base metal sulphidation potential and to the right of the kinetic boundary, favour oxide scale formation. The scale thickness varied somewhat, but the most important differences between the alloy

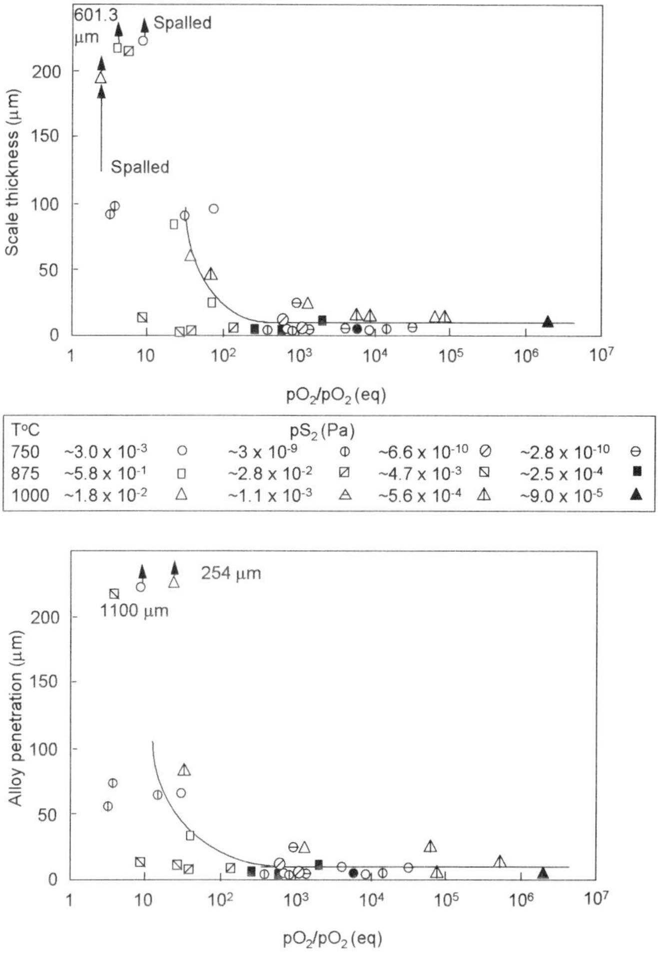

Figure 7.10 Variation in scale thickness and alloy penetration as a function of excess oxygen parameter for Alloy 800 exposed to oxygen/sulphur mixed gas environments.

types pertain to the subscale structure and the shape of the metal/oxide interface. *Régime 3* conditions, above the base-metal sulphidation potential and to the left of the kinetic boundary, result in scales that were complex mixtures of the base-metal sulphide, chromium sulphide, Cr-depleted alloy, and the appropriate oxide. Accelerated liquid phase corrosion occurred if nickel was in sufficient supply in an alloy to form nickel-rich sulphide. A schematic reaction for attack in *Régime 3* is shown in Fig. 7.12. Competition between oxide and base-metal sulphide nuclei is followed by sulphide overgrowth, which has a considerably higher growth rate than that of the oxide. Void formation in the subscale occurs throughout the entire sequence, because of outward cation diffusion. Chromium sulphide forms at the sulphide/metal interface through the displacement reaction between the

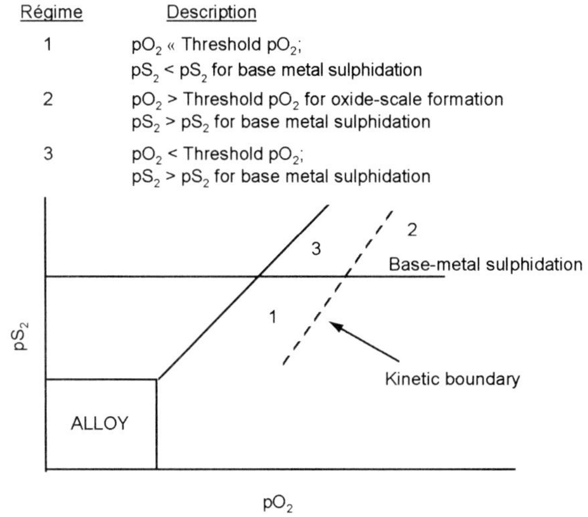

Figure 7.11 Schematic of oxygen/sulphur régimes of material behaviour.

base-metal sulphide and the chromium diffusing from the matrix. Finally, the sulphides dissociate at the heavily voided interface, and the sulphur released into the voids diffuses into the matrix, forming internal chromium sulphide.

Because the type of corrosion product scale that forms on an alloy exposed to a mixed gas atmosphere is determined by the relative rates of oxidation versus sulphidation of the scale-forming elements, any improvement in physical characteristics (alloy grain size, surface finish) and chemical characteristics (alloy composition, alloying element distribution) that can enable nucleation/growth of oxide in the early stages of alloy exposure will be beneficial. In this vein, *oxide dispersions* and *refractory element additions* have been found beneficial in improving the sulphur resistance of alloys. For example, Fig. 7.13 shows the oxidation/sulphidation behaviour of several commercial alloys and Fe–Cr–Al and Fe–Cr–Ni–Al alloys with oxide dispersions [11]. Even though the scale compositions are essentially similar in all these alloys, the scaling rates are orders of magnitude lower in the alloys with dispersion of rare earth oxide particles.

Refractory metals such as Ti, Zr, V and Ta can form oxides and sulphides that exhibit greater thermodynamic stability than do chromium oxide and sulphide. Figure 7.14 shows a comparison of thermogravimetric test data for several refractory metals and alloys with data for conventional alloys exposed to mixed gas atmospheres [12]. Even though the refractory metals and alloys exhibit low corrosion rates when exposed to complex atmospheres, use of these materials as structural materials in fossil technologies is *not* feasible because of difficulty in fabrication, inadequate mechanical properties, and cost. However, the refractory metals can be used as *alloying additions* to existing structural alloys for improved corrosion resistance.

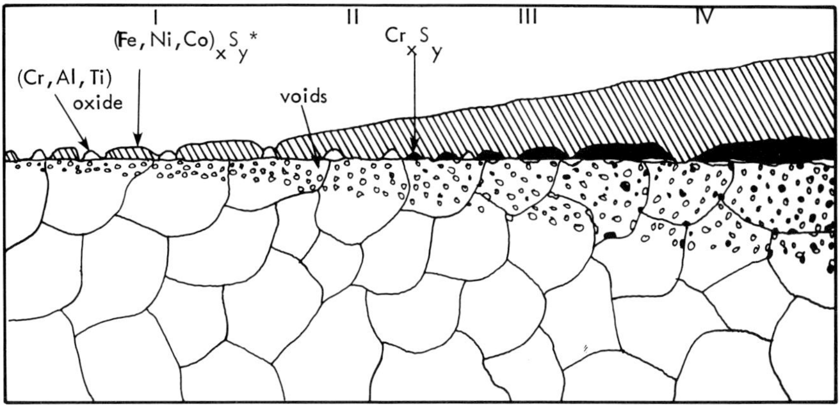

I Competition
II Overgrowth
III Cr_XS_Y formation
IV Dissociation & internal sulphidation

* If the base metal sulphide is molten at test temperature, islands of Cr_XS_Y, oxides and Cr-depleted alloy exist in the outer scale after cooling due to grain boundary attack by the liquid.

Figure 7.12 Schematic reaction sequence for corrosion of chromia-forming alloys in *Régime 3*.

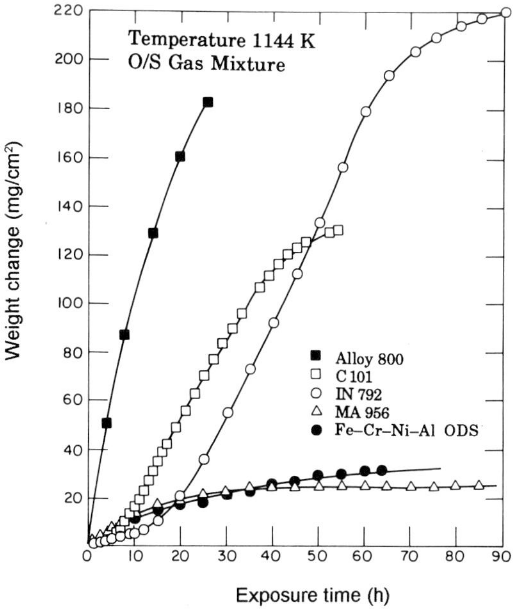

Figure 7.13 Comparison of thermogravimetric test data for several alloys tested at 871°C in oxygen/sulphur mixed gas.

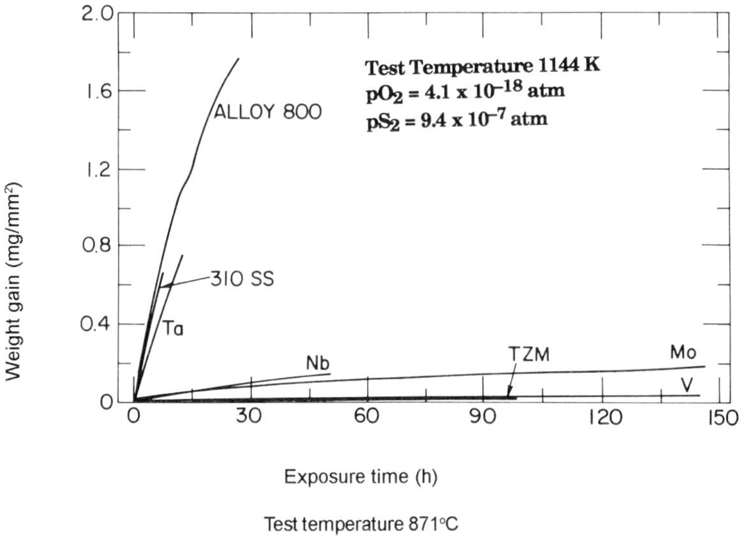

Figure 7.14 A comparison of thermogravimetric test data for several refractory metals/ alloys with those of conventional alloys exposed to oxygen ($pO_2 \sim 4.1 \times 10^{-13}$ Pa) sulphur ($pS_2 \sim 9.4 \times 10^{-2}$ Pa) mixed gas atmospheres.

Extensive corrosion studies have been conducted on iron-base alloys with and without additions of Nb and Zr. Detailed analyses of gas/scale and scale/metal interfaces indicate that the additions of Nb or Zr minimize or eliminate the voids at the scale/metal interface, thereby improving the adhesion of the chromia scale to the substrate. Figures 7.15 and 7.16 show scanning electron micrographs of the specimens (without and with Zr additions, respectively) that were tested at 875°C in oxygen/sulphur environments [12]. Further additions of Nb and Zr lead to formation of a continuous layer of refractory metal oxide at the interface between the chromia and the alloy substrate. This inner layer acts as a *barrier* to outward transport of cations from the alloy and inward transport of sulphur from the gas phase, thereby minimizing sulphidation of base-metal sulphides and prolonging the time before onset of *breakaway corrosion*.

7.4.3 Behaviour of Al_2O_3 alloys

The only thermodynamically stable solid oxide in the Al-O system is Al_2O_3, which has a melting point of 2072°C. Various structural forms of the oxide are possible, by only *a*-alumina, with a corundum structure like that of Cr_2O_3, is important in high temperature corrosion. Cr_2O_3 normally exhibits more complex electrical properties whereby *either* Al *or* O may be mobile. Based on bulk self diffusion data, the rate of Cr_2O_3 scale formation would be expected to exceed the rate of

Al_2O_3 scale formation by more than six orders of magnitude at temperatures <1177°C, assuming growth to be controlled by cation diffusion in each case. However, from a review [13] of the kinetics of Cr_2O_3 and Al_2O_3 scale growth, it was concluded that for approximately parabolic kinetics, the rate constant for Cr_2O_3 formation may exceed that for Al_2O_3 by only two orders of magnitude for similar conditions of exposure. Oxygen partial pressure in the exposure environment appears to have little influence on the rate of Al_2O_3 scale formation [14], but impurities may dope the oxide and affect its growth characteristics [15]. The rate of sulphur diffusion through Al_2O_3 can be one to four orders of magnitude greater than the rate of oxygen diffusion, depending on oxide grain size, but may

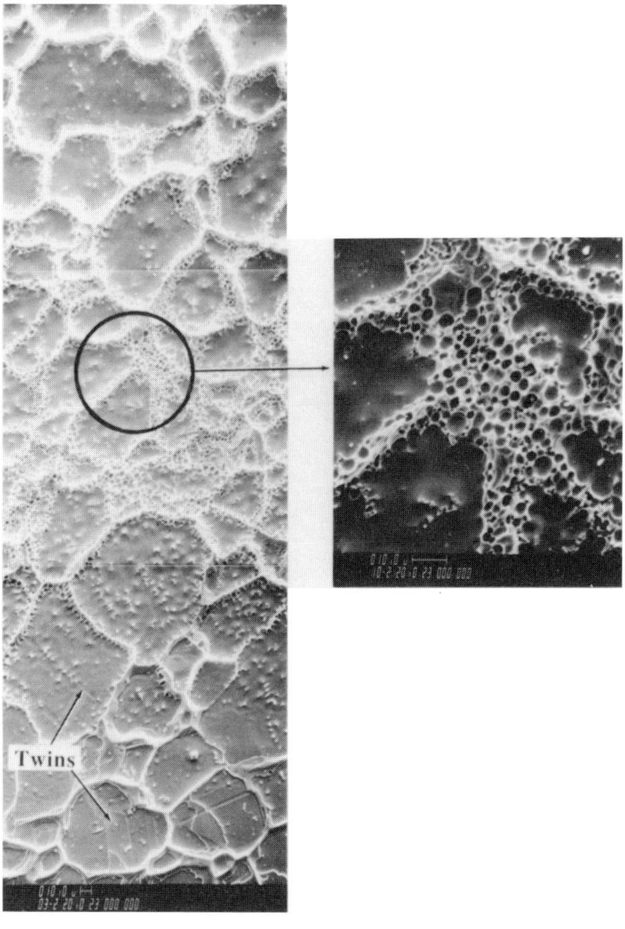

Figure 7.15 Scanning electron microscopy photographs at different magnifications of metal side of scale/metal interface on Fe25Cr20Ni ternary alloy after exposure to low oxygen, high sulphur environment.

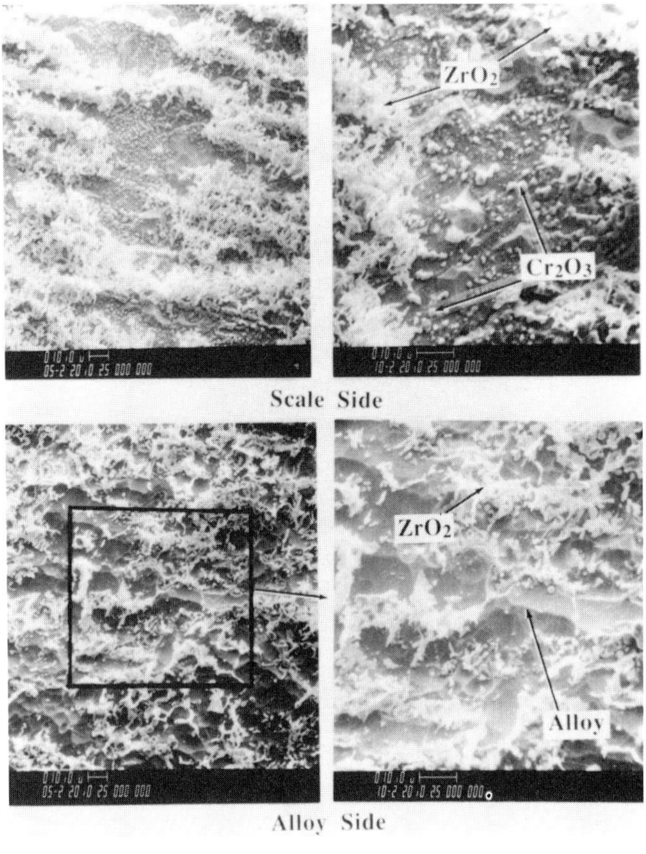

Figure 7.16 Scanning electron microscopy photographs at different magnifications of scale and metal sides of scale/metal interface on Fe25Cr20Ni3Zr alloy after exposure to low oxygen, high sulphur environment.

be 1.5 to 4.5 orders of magnitude lower than the rate of sulphur diffusion through Cr_2O_3. In Ni–Al, Fe–Al and Co–Al binary alloys, an Al content in excess of 10 wt.% is normally needed to support formation of a surface Al_2O_3 scale [16]. Alumina formation is enhanced by an increase in temperature [17]. The presence of Cr reduces the required amount of Al to 3 to 5 wt.% for the formation of Al_2O_3. In Fe–Cr–Ni–Al alloys, a combination of good mechanical properties and oxidation resistance is obtained by optimising the alloy chemistry. A Cr content of 20 wt.% or more is necessary to support Al_2O_3 scale formation with only 5 wt.% Al; more than 5 wt.% Al *reduces alloy workability* [18]. In structural alloys with an Al content of 3 to 5 wt.%, there is generally an insufficient quantity of Al to support repeated scale repair operations; as a result, considerable effort has been expended to improve the mechanical stability of Al_2O_3 scales.

In general sulphidation of Al-containing alloys proceeds at a rate that is orders of magnitude greater than the rate of oxidation [19]. In mixed oxygen- and sulphur-containing environments where an Al_2O_3 scale can be formed, the rate of scale growth may be similar to that observed for pure oxidation, but relatively small changes in partial pressures of oxygen and sulphur can abruptly result in accelerated corrosion. For example, Huang et al [20] have shown that the parabolic rate of constant for scaling an Fe18Cr6Al alloy at 900°C can increase from $8.2 \times 10^{-4} g^2 cm^{-4} s^{-1}$ in a pure oxygen atmosphere to $7.1 \times 10^{-3} g^2 cm^{-4} s^{-1}$ in a gas mixture with $pO_2 \sim 4 \times 10^{-15}$ and $pS_2 \sim 1 \times 10^{-3}$ Pa. Furthermore, the results of marker experiments in this study suggest that the presence of sulphur in the environment alters the scale growth mechanism from predominant inward oxygen diffusion to predominant outward diffusion of aluminium cations. Under these conditions, continued oxide growth seems to occur at the scale/gas interface; such a growth process has been suggested as the cause for enhanced *scale adherence* evidenced by the lack of craters on the bare metal surface and by the lower amount of *acoustic emission* (related to cracking propensity) in the scales grown in sulphur-containing environments. In contrast with high Cr alloys, which exhibit a threshold pO_2 for Cr_2O_3 formation in mixed gas atmospheres, the aluminium-containing alloys generally form stable Al_2O_3 scales upon exposure to oxygen/sulphur mixed oxidant atmospheres; however, the long term stability of the scales in terms of cracking, spallation, and lack of adhesion is of concern.

7.4.4 Behaviour of SiO_2-forming alloys

Very limited information is available on the oxidation behaviour of silica-forming alloys. Even though SiO_2 is thermodynamically more stable than Cr_2O_3, the mobility of silicon is much lower than that of Fe or Cr in Fe- and Ni-base alloys; therefore, an external SiO_2 scale rarely forms on those alloys via oxidation of the substrate silicon [21, 22]. Preliminary studies on Fe-Si and Ni-Si binary alloys exposed to oxygen/sulphur environments indicate that silicon levels in excess of 10 wt.% will be needed for silica (as an external scale) formation; such high concentrations make the alloy *too brittle* (mechanically) for use as structural alloys [23].

7.4.5 Breakdown of scales in mixed gas environments

The mechanism by which the corrosion process proceeds in a given alloy is strongly dependent on both the alloy chemistry and the partial pressures of the reactants in the gas phase. For example, in the oxygen/sulphur bioxidant studies, under *Régime 3* conditions mentioned above for Cr_2O_3-forming alloys, competition between oxide base metal sulphide determines the type of scale that forms on a metal surface. On the other hand, under *Régime 2* conditions, the same alloys develop oxide scales. The *threshold* pO_2 values for oxide formation are temperature-dependent, but are influenced little by the chromium content of the alloy in the range of 20–50 wt.%. Minor alloying additions primarily influence the type (binary oxides, spinels and duplex layers) and porosity of the scales as well as the adhesion of the oxide scale to the substrate material. Even if an alloy develops a protective oxide scale after short term exposures to mixed gas environments, the long term behaviour and thus life expectation for the alloy is strongly dependent on whether the alloy exhibits "*breakaway*" or "*accelerated*" corrosion.

Several possible *breakaway* processes can be identified, including

- Mechanical and thermal cycling damage.
- Development of short-circuit and impurity transport paths in the oxide scale.
- Changes in the oxide composition with time.
- Depletion of the protective scale-forming element in the substrate as a result of repeated spalling and reforming of the protective scale.
- Transport of base metal cations through the oxide scale to the gas/scale interface and subsequent reaction of these elements with the second reactant in the gas phase.
- Transport of second reactant through the oxide scale into the substrate and reaction of the chromium- or aluminium-depleted (Fe, Ni, Co) region in the vicinity of the scale/metal interface.

In general, most of the alloys exhibit *breakaway* corrosion; especially in oxygen/sulphur mixed gas atmospheres; the exposure time at which it occurs is dependent on temperature, gas chemistry, alloy composition, and microstructure of scales. In most applications of heat resistant alloys, the materials are subjected to temperature cycling conditions. *Breakdown* of scales can occur due to the difference in thermal properties between scales and alloy substrates and to growth stresses generated during oxidation. Baxter and Natesan [24] have discussed various mechanical considerations in the degradation of structural materials exposed to several environments at elevated temperatures. In addition, transport of substrate elements through the oxide scale to the gas/scale interface and subsequent reaction with the second oxidant can occur during bioxidant exposure. In general, the diffusivities of various elements in decreasing order are $D_{Mn} > D_{Fe} > D_{Cr} > D_{Co} > D_{Ni}$. As a result, even among chromia-forming alloys, high iron and manganese alloys are much more susceptible to reaction with the second oxidant than are the cobalt- and nickel-base alloys; however, the second oxidant can and does transport through the Cr_2O_3 scales and reacts with substrate elements.

Although the onset of *breakaway corrosion* is difficult to predict, the influence of alloy and gas chemistries on the catastrophic nature of the corrosion process is of concern in the application of materials at elevated temperatures. The scale morphologies that develop on alloys exposed to mixed gas atmospheres are strongly dependent on the amount of second reactant present in the gas phase. Nucleation and subsequent growth of tenacious protective oxide scales can be affected by the adsorption onto, and transport of the second reactant through, the scale layers. Further, a knowledge of time-dependent variations in the composition of scale layers, in developing new reaction product phases as well as changes in the physical properties of scale, is needed to establish the susceptibility of an alloy to *breakaway corrosion*.

Based on available information on the oxidation/sulphidation behaviour of structural alloys, one can categorize the alloys in terms of two distinct corrosion scale (depending of whether sulphur is present as H_2S or SO_2) morphologies. Figure 7.17 shows the surface of an Fe–Cr–Ni–Zr specimen as the scaling process advances with time upon exposure to H_2S-containing mixed gas environment. Figure 7.18 shows the scale morphology in cross-sections of specimens of Fe25Cr20Ni alloy that have been preoxidized for 72 h at 875°C in a sulphur-free atmosphere with $pO_2 \sim 2 \times 10^{-13}$ Pa and subsequently exposed to an H_2–H_2S atmosphere ($pS_2 \sim 4 \times 10^{-3}$ Pa) for 5,7 and 22 h. Preoxidation of the alloy in a low pO_2 environment resulted in an external Cr_2O_3 scale 4 μm thick. After 5h exposure

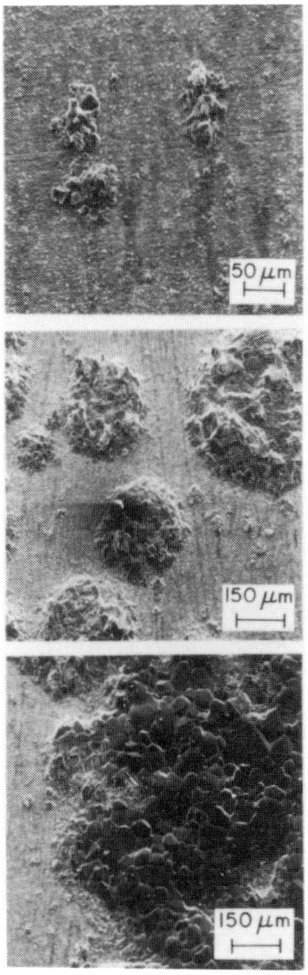

1) Particles of Fe, Cr sulphide appear on the surface of the oxide scale.

2) Sulphide particles grow into islands of sulphide becoming richer in Fe.

3) Sulphide islands eventually impinge and the composition approaches that of iron sulphide.

Figure 7.17 Stages of sulphide scale development on Fe12Cr12Ni (0–6) Zr alloys in oxygen/sulphur mixed gas at 875°C.

(a)

(b)

(c)

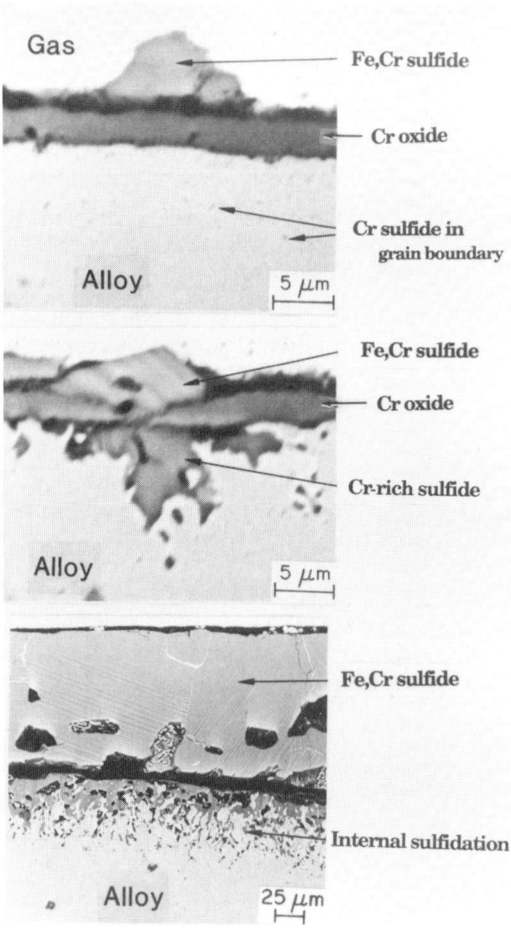

Fe 25Cr20Ni Preoxidised for 72 h in pO2 ~ 2 x 10^{-13} Pa
sulphidised in pS2 ~ 4 x 10^{-3} Pa for (a) 5h; (b) 7h; (c) 22h.

Figure 7.18 Morphological changes observed in Fe25Cr20Ni alloy preoxidised in low pO_2 atmosphere and subsequently exposed to H_2 - H_2S atmosphere for different times at 875°C.

to the H_2–H_2S atmosphere, the preoxidized specimen developed an (Fe, Cr) sulphide phase at the gas/oxide interface, indicating significant transport of Fe and Cr through the oxide scale and subsequent sulphidation of the transported elements. Small precipitates of Cr-rich sulphide particles can also be observed at the grain boundaries in the substrate material, indicating some inward transport of sulphur. After a 7h exposure, the continuous oxide scale has been breached, and sulphidation at the substrate/oxide interface is noted. The sulphide particles at the grain boundaries become larger due to increased sulphur penetration. After

22 h of exposure, the oxide scale is virtually destroyed, and the scale consists predominately of (Fe, Cr) sulphide accompanied by substantial internal sulphidation of the alloy.

The morphological features that develop in high chromium alloys exposed to O_2–SO_2 mixed gas environments are somewhat different when compared with those obtained in an H_2–H_2S environment described above. In general, the O_2–SO_2 gas mixtures establish much higher oxygen partial pressure in the gas phase (than those of H_2–H_2O–H_2S mixtures); as a result, the alloys exposed to such environments develop predominantly oxide scales. However, the oxide scales are porous in nature and *also* generally contain a sulphide phase that results from the reaction between the base-metal constituents and the sulphur released from the oxidation of Cr or Al with SO_2. The porous oxide scale enables SO_2 gas

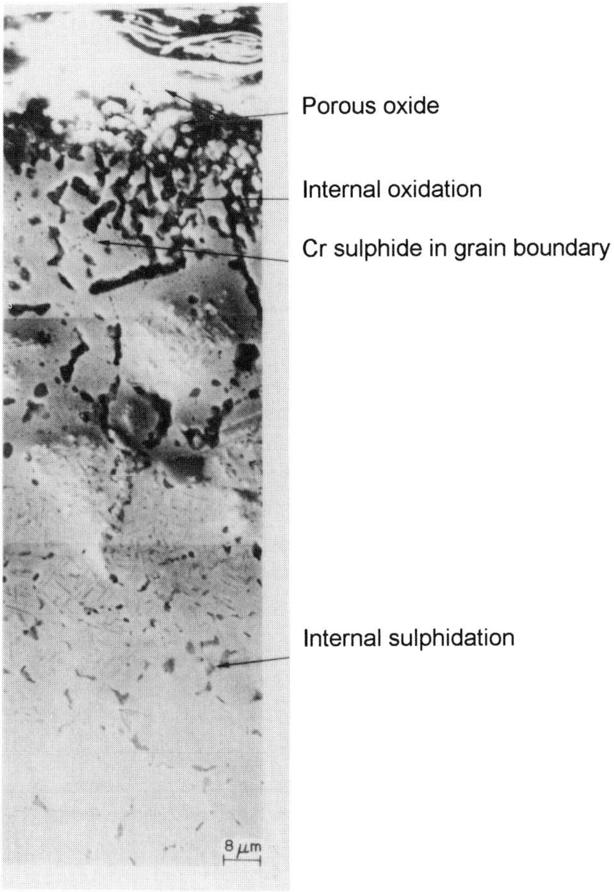

Figure 7.19 Morphological changes observed in Alloy 800 specimen after 500 h exposure to O_2–SO_2 atmosphere at 840°C.

molecules to permeate to the scale/substrate interface and cause internal sulphidation. Figure 7.19 shows the morphological features that developed on Alloy 800 after a 500h exposure at 900°C to O_2–SO_2 gas mixture with $pO_2 \sim 7 \times 10^2$ and $pSO_2 \sim 1.7 \times 10^2$ Pa. The sulphur partial pressure corresponding to this O_2–SO_2 gas mixture is $\sim 2.6 \times 10^{-23}$ Pa, which is extremely low for sulphidation of chromium or any of the other elements in the alloy. However, the pSO_2 value is orders of magnitude higher than that required for sulphidation of Cr and Fe in the alloy, and the internal sulphidation of the substrate material occurs via reaction between SO_2 gas transported through the porous oxide and the substrate elements. In such environments, deep internal sulphidation of the material prevents the alloy from developing a protective oxide scale and eventually leads to *breakaway corrosion*.

Figures 7.20 and 7.21 are schematic representations of corrosion scale development and morphological changes that occur in Cr_2O_3 -forming alloys exposed to low pO_2 (with H_2S) and high pO_2 (with SO_2) atmospheres, respectively, at elevated temperatures. In the former atmosphere (under *Régime 2* conditions discussed earlier), the alloy in the early stages of exposure develops oxide *and* sulphide nuclei. Eventually, the thermodynamic conditions establish a continuous Cr_2O_3 scale via reoxidation of sulphide particles, while the sulphur released is driven into the substrate along the grain boundaries. At the same time, the sulphur in the gas phase is adsorbed onto the scale/gas interface, and channels are established in the fine grained oxide scale through the transport of base-metal cations (Cr, Fe, Ni, Co, etc.) to the scale/gas interface is accentuated. If sulphur pressure in the gas phase exceeds the metal/metal sulphide equilibria for the base-metal elements, their sulphides are formed at the oxide scale/gas interface. As the sulphide grows, stresses develop in the oxide scale, which eventually is breached and leads to sulphidation at the oxide scale/alloy interface. Because the transport rates of cations and sulphur through the sulphide phase are orders of magnitude faster than those through the oxide scale, the sulphidation attack continues in an accelerated manner. At longer exposure times, the oxide is virtually destroyed, and a massive sulphide scale develops – a condition that represents *breakaway corrosion* for the alloy. The same sequence of steps is operative in alumina-forming alloys, although probably at a much slower rate than in chromia-forming materials.

In the case of high pO_2 (with SO_2) atmospheres, the high chromium alloys generally develop porous oxide scales, and some sulphides are also observed in the inner portions of the scale (in the vicinity of scale/substrate interface). The sulphides form via reactions between substrate elements and sulphur that is released when chromium reacts with SO_2 to form the external oxide scale. The porosity present in the scale enables further molecular transport of SO_2 in the gas phase to the scale/substrate interface, leading to further oxidation/sulphidation. The sulphur that is released is transported along the grain boundaries in the metal, leading to internal sulphidation of the alloy. Generally, the acceptable lifetimes for high Cr

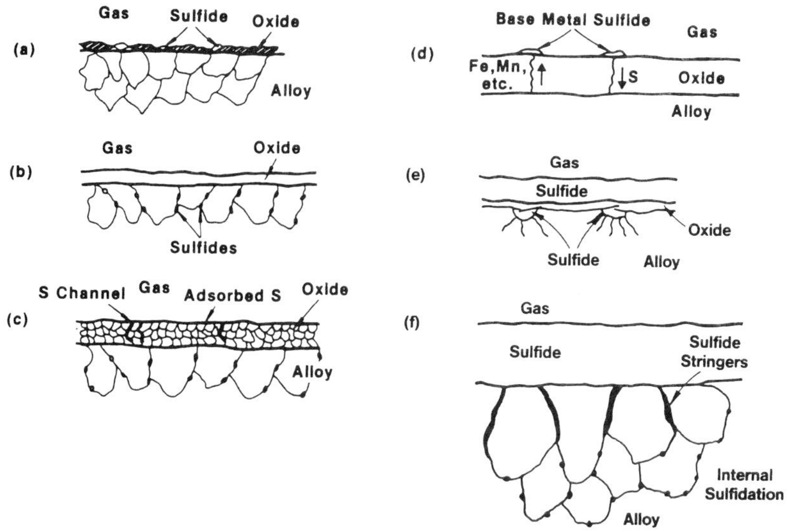

Figure 7.20 Schematic representation of reaction sequence for chromia-forming alloys exposed to low pO_2 environments containing H_2S.

alloys exposed to these atmospheres are determined by the magnitude of the depth of internal sulphidation, which is largely determined by the alloy chemistry, temperature, and SO_2 content of the gas phase.

The *breakdown* of scales in alumina-forming alloys has been studied to a lesser extent [20,25], but three modes of failure can be identified: (a) the *spallation* of the scale because of thermal cycling and inability to reform the protective scale while exposing the base-metal to sulphidation attack; (b) *transport of base-metal elements* to the scale/gas interface (but at a slower rate than in Cr_2O_3-forming alloys) and subsequent sulphidation for these elements at the gas/scale interface; and (c) *enhanced transport of Al* to the scale/gas interface and formation of

Table 7.1 Typical chemical composition in (wt%) of candidate refractories

	Castables			**Shapes**		
	Bubble Al_2O_3	**Dense Al_2O_3**	**SiC**	**Al_2O_3**	**Fused cast Al_2O_3**	**SiC**
Al_2O_3	94.6	96.0	6.6	99.2	98.7	0.7
SiO_2	0.5	0.1	2.2	0.5	0.5	8.5
Fe_2O_3	0.2	0.1	1.8	0.1	0.1	0.7
CaO	4.2	3.6	5.5	0.1	0.1	0.2
Alkalis	0.4	0.4	-	0.1	0.04	-
Other	-	-	82.7 (SiC)	-	0.5 (B_2O_3)	89.6 (SiC)

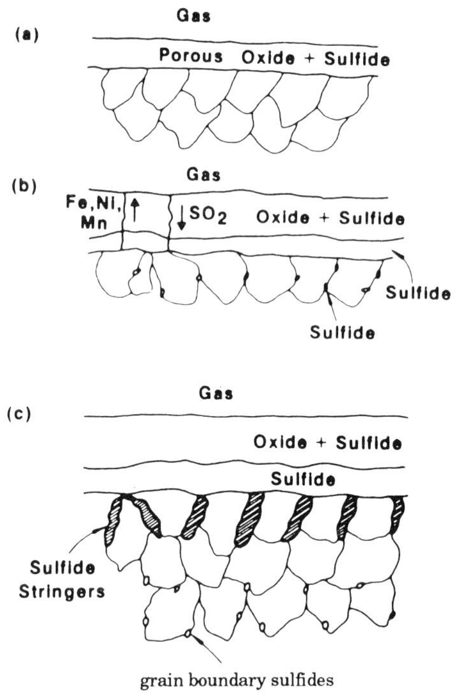

Figure 7.21 Schematic representation of reaction sequences for chromia-forming alloys exposed to high pO_2 environments containing SO_2.

aluminium sulphide particles in the Al_2O_3 scale in the vicinity of the gas/scale interface.

7.5 Gaseous corrosion of refractory materials

Refractory linings and components are being considered for use in several process schemes where metallic materials do not perform adequately or where service temperatures are very high for reliable application of metallic materials. Refractories are complex multiphase materials comprising refractory grains or aggregates bonded within sintered, fused or cement matrices. *Fired or fused refractory* shapes have glassy or crystalline matrices while *castable (concrete) refractories* have hydraulic or chemically bonded matrices with significant porosity. Many of the refractories used in gaseous corrosion environments are either of the high alumina type (where severe erosion conditions exist) or SiC refractory shapes or castables (where high thermal conductivity is desirable). Chemical compositions of several candidate refractories are given in Table 7.1 [26].

The forms of gaseous reaction of refractory materials include reactions with steam, hydrogen, carbon and alkali vapours. Dry hydrogen or water vapour in a gas can attack and remove silica from refractories. Dry hydrogen will reduce solid SiO_2 to gaseous SiO at temperatures above 927°C. Further, steam can remove silica as volatile silicic acids at temperatures as low as 1050°C under pressure. Removal of SiO_2 by either mechanism affects the refractory strength and erosion resistance. Because most gas environments contain steam, distillation of the silica may be a primary mechanism of attack on the refractories. However, the rate of silica removal may not be linear with time so that only the surface layers undergo high silica removal even though reactions may proceed behind porous refractory linings.

To prevent possible failure by silica removal through reactions with steam and/or hydrogen, mostly low silica refractories can be used in the linings exposed to the gaseous environment. Low silica content is especially important for the bonding phases where weakening can result in refractory failure independently of aggregate or grain. Based on this and the reported low rate of silica loss in the *ammonia secondary reformer* application, dense castables consisting of high fired dense fireclay aggregates bonded with high purity calcium aluminate cements may be satisfactory for the low temperature linings. Dense high fired high alumina fireclay brick may also prove satisfactory for this application.

Phosphate bonded monoliths can undergo weakening in reducing atmospheres and steam at high temperatures because the phosphate bond may be chemically attacked and dissolved [26]. At very high temperatures under reducing conditions, the phosphate bond may be reduced and phosphorus *vaporized*. Phosphate bonded high alumina refractories fired at high temperatures to develop ceramic bonds should be less susceptible to failure under reducing conditions and steam. This type of refractory is used for the erosion/corrosion conditions in some systems, while phosphate-bonded castables are frequently used for repairs.

Carbon monoxide disintegration of refractories containing free iron, iron oxide, or other iron-bearing compounds occurs via the reaction $2CO = CO_2 + C$. This reaction generally occurs at 377 to 677°C with the carbon depositing in the vicinity of the iron in the refractory. As the deposited carbon builds up, high local stresses develop until cracking and disintegration occur, usually in the form of spalling of the refractory. The reaction is favoured at high CO and low CO_2 pressures. The carbon deposition may actually occur behind the hot face of the refractory lining, leading to spallation of entire sections of the hot face.

7.6 Deposit-induced corrosion

Corrosion of materials in the presence of deposits of various types is a concern in materials selection and in their adequacy of performance when used in several of the process environments. Under practical usage of material, corrosion becomes fairly complex in view of interactions between the deposit constituents and the

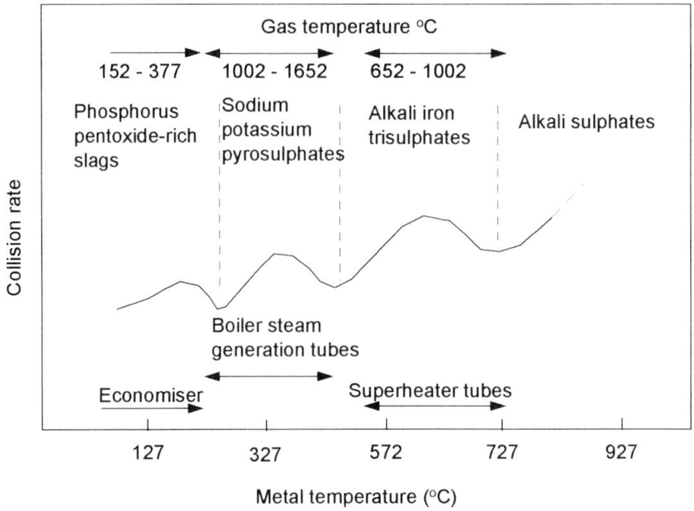

Figure 7.22 Régimes of fireside corrosion in coal-fired boilers.

elements/phases of the substrate materials, wide variations in the local chemistries in the vicinity of the zone of interaction, time-dependent variations in chemistries in the deposit and the substrate materials, and influence of corrosion product layers in corrosion kinetics.

7.6.1 Corrosion due to deposits containing alkali, sulphur and oxygen

Fireside metal wastage in conventional coal-fired boilers can occur via gas phase oxidation or deposit-induced liquid phase corrosion. The former can be minimized by using materials that are oxidation resistant at service temperatures of interest. On the other hand, *deposit-induced corrosion* of materials is an accelerated type of attack influenced by the vaporization and condensation of small amounts of impurities such as sodium, potassium, sulphur, chlorine, and vanadium, or their compounds, present in the coal feedstock. The effect of boiler deposits on the corrosion of structural materials, has been fairly well established by Reid [27] and Wright et al [28], and the temperature régimes in which this corrosion occurs are summarized in Fig. 7.22. The data generally show that $K_2S_2O_7$ will form K_2SO_4 and SO_3 at 400°C when SO_3 concentration is at least 150ppm; as the temperature increases, the SO_3 requirement increases, so that at ~500°C at least 2000 ppm SO_3 will be required to form liquid $K_2S_2O_7$. Sodium pyrosulphate can form at ~400°C with about 2500 ppm SO_3, but
2 vol.% SO_3 will be required at 480°C. Based on these results and the anticipated maximum level of ~3500 ppm SO_3 in a *pulverized-coal boiler*, Reid [27] concluded that pyrosulphates can contribute to metal loss in the *waterwall* and *economizer tubes* but may not be a cause of corrosion in *superheaters* and *reheaters*.

Because boiler tube materials are iron-base alloys, and because mobility of

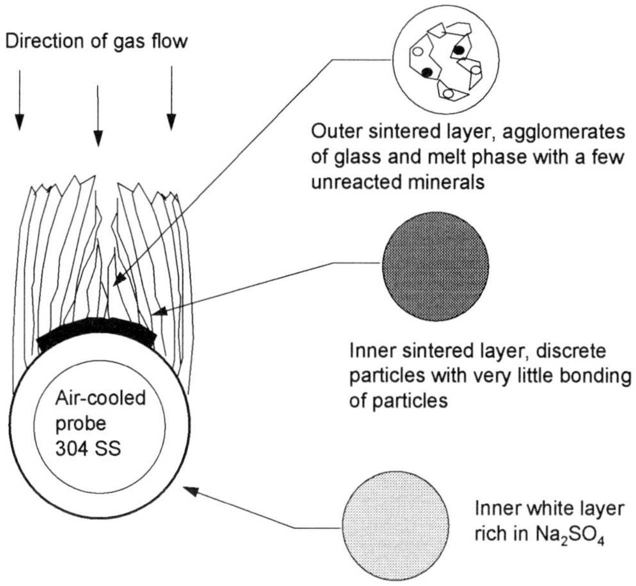

Figure 7.23 Deposit structure on superheater tubes in pulverised-coal-burning boilers.

iron from the alloy substrate to the scale/gas interface is fairly rapid, considerable attention has been given to understanding the formation of alkali iron trisulphates and their rôle in the corrosion of steam superheaters [27,29,30]. Figure 7.23 shows the deposit characteristics in superheater sections that can lead to fouling and corrosion of the underlying metal. A general mechanism and sequence of events for *alkali metal-induced corrosion* of iron-based materials have been described by Reid [27]. Initially, an oxide film forms on the metal surface; also, pyrite in the coal can oxidize during combustion to form iron oxide and sulphur dioxide/sulphur/ trioxide gases. Alkali sulphates, originating from alkalis in the coal and sulphur oxides in the surface atmosphere, are deposited over the oxide scale on the superheater material. Eventually, the outer surface of the alkali sulphate layer becomes sticky because of the increasing temperature gradient, so that particles of fly ash are captured. With further increase in temperature, thermal dissociation of sulphur compounds in the ash releases SO_3 that migrates toward the cooler metal surface, while a layer of slag forms on the outer surface. With more ash in the outer layer, the temperature falls in the sulphate layer, and reaction between the oxide scale and SO_3 forms alkali iron trisulphate. With this removal of the oxide scale, the metal oxidizes further. Deslagging due to temperature excursions or soot blowing to remove the deposit exposes the alkali-iron trisulphates to higher temperatures and results in dissociation of sulphate and generation of SO_3 for further attack of the metal.

A number of factors, including sulphur, alkali, chlorine in coal feedstock, excess air level used in the combustion process, and metal temperature, determine the

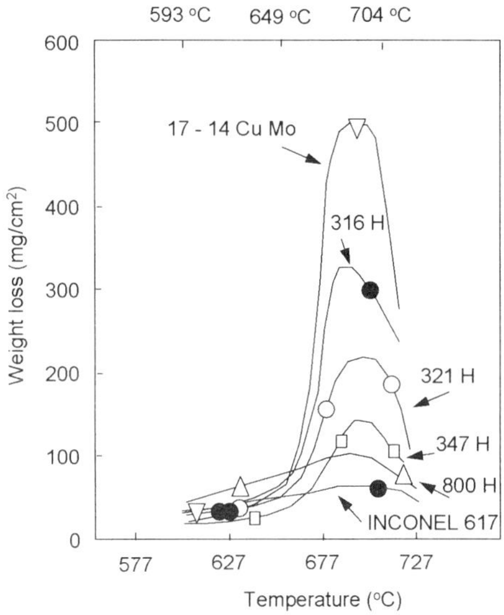

Figure 7.24 Corrosion test data for several alloys exposed to coal-fired boiler environments.

extent of corrosion of superheater materials in coal-fired boilers. Typical corrosion results [31] for several candidate alloys are shown in Fig. 7.24. It is evident that at a *steam temperature* of ~ 565°C the *metal temperature* will be in the range 600-625°C and the corrosion rates will be acceptable for long-term service. With advanced steam cycles in which steam temperatures and pressures of 650°C and 34.5 MPa are anticipated, the metal temperature can attain 700°C or higher, resulting in increased corrosion rates. It is obvious that new materials or corrosion protection of existing materials are needed for reliable service of superheaters in advanced steam cycle plants.

7.6.2 Corrosion in the presence of calcium-containing deposits

The fluidized bed combustion (FBC) of coal produces gas that principally contains O_2, CO_2, H_2O, SO_2 and N_2, together with minor amounts of SO_3, nitrogen oxides, chlorides, and other volatilized salts. The gas composition is highly dependent on the air/coal stoichiometric ratio. In addition, SO_2 concentration in the gas phase will be determined by the type and amount of sulphur sorbent and sulphur content of the coal used in the combustion process; however, the local chemistry beneath the deposit could be quite reducing relative to the bulk gas composition. The oxygen/sulphur thermochemical diagrams shown in Fig 7.25 for Fe, Cr and Ni at several temperatures indicate the thermodynamic stability of various oxide and sulphide phases. Superimposed on these diagrams are the CaO–CaS–$CaSO_4$ phase fields at the corresponding metal temperatures. The extent of interaction between

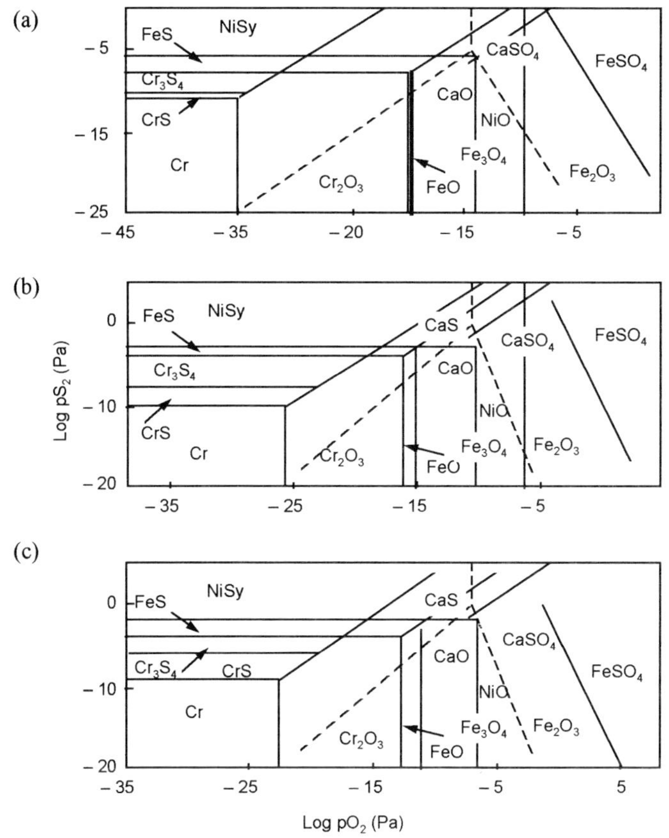

Figure 7.25 Oxygen/sulphur thermochemical diagrams at (a) 593, (b) 977 and (c) 840 °C, depicting regions of stability of various oxide and sulphide phases. Phase fields for CaO–CaS–$CaSO_4$ system (dashed lines) are also shown.

the deposit and the substrate or the deposit and the scale depends on three factors: (a) porosity of the deposit layer and the *transport* of gaseous molecules containing sulphur; (b) *dissociation* of $CaSO_4$, to establish a sulphur pressure at the underside of the deposit; and (c) *rate of reaction* between the underlying alloy elements and the reactants, such as oxygen and sulphur, to form oxide/sulphide scales and enable penetration of oxygen/sulphur into the substrate.

Experiments have been conducted to examine the combined effect of sorbent/gas chemistry on the corrosion of structural alloys over a wide range of metal temperatures, gas chemistries, gas cyclical conditions, and deposit types [32, 33]. Figure 7.26 shows scanning electron microscopy (SEM) photographs of corrosion product layers that developed on Alloy 800 specimens, coated with deposit mixtures (defined by *lines 1-3* and *point 4* in the pO_2-pS_2 diagram Fig. 7.26), after exposure at 840°C to a gas mixture with oxygen and sulphur partial pressures ~ 5.4 x 10^{-7} and ~ 1.6 x 10^{-3} Pa, respectively. The four synthetic

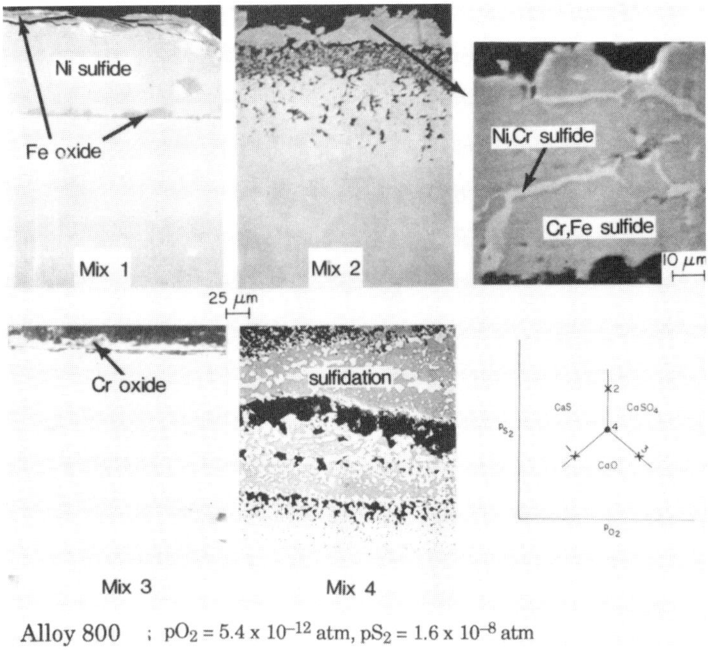

Figure 7.26 Morphologies of scale layers that developed on Alloy 800 specimens coated with various deposit mixtures (defined by lines 1–3 and point 4 of the $CaO–CaS–CaSO_4$ phase stability diagram) and exposed to gas mixture with pO_2 and pS_2 of ~ 5.4×10^{-7} and ~ 1.6×10^{-3} Pa, respectively.

deposit mixtures lead to oxygen and sulphur partial pressures dictated by the following chemical reactions:

$$CaO + \frac{1}{2}S_2 = CaS + \frac{1}{2}O_2 \qquad \textit{Mixture 1}$$

$$CaS + 2\,O_2 = CaSO_4 \qquad \textit{Mixture 2}$$

$$CaO + \frac{1}{2}S_2 + \frac{3}{2}O_2 = CaSO_4 \qquad \textit{Mixture 3}$$

$$CaO + CaS + \frac{1}{2}S_2 + \frac{7}{2}O_2 = 2CaSO_4 \qquad \textit{Mixture 4}$$

Specimens coated with *mixtures 1, 2, and 4* exhibited sulphidation attack in these experiments, whereas the specimen coated with *mixture 3* still developed a thin chromium oxide scale. The implications of these tests is that when *mixtures 2 and 4* are in the presence of SO_2 in the gas phase, a fairly high pS_2 and low pO_2 can be established in the pores of the deposit material and in the deposit/alloy interface region. Consequently, the alloy has a tendency to undergo sulphidation attack in these two tests. In the presence of *mixture 1* and SO_2 in the gas phase, the dominant reaction in the deposit will be sulphidation of CaO, which will result in an increase in pO_2 and pS_2 in the pores of the deposit; however, the pO_2 will still be below that dictated by CaO/CaS/$CaSO_4$ triple point. As a result, the alloy will undergo sulphidation attack, and nickel sulphide and iron oxide will be the reaction product phases. In the presence of *mixture 3* and SO_2 in the gas phase, the dominant reaction in the deposit will be the sulphation of CaO to $CaSO_4$ (which can decrease the pS_2 in the pores of the deposit material) and the alloy tends to undergo oxidation.

Pretreatment of Specimens The rôle of two pretreatments (*preoxidation* and *precarburization*) of alloy specimens in the subsequent corrosion of alloys in the presence of different deposit mixtures and gas environments has been examined. In *preoxidation* treatment, the alloys were oxidized for ~ 80 h at 840°C in a sulphur-free gas atmosphere with pO_2 of ~ 3.6×10^{-9} Pa. In the *precarburization* treatment, the specimens were carburized at 840°C in a 5 vol%. CH_4–H_2 gas mixture for 64 h.

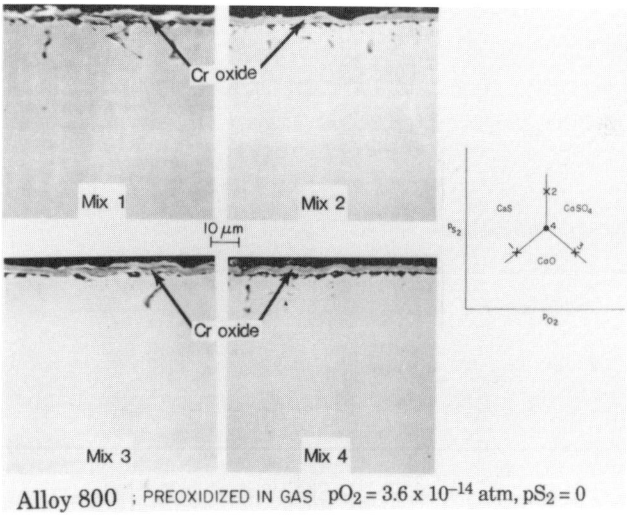

Figure 7.27 Morphologies of scale layers that developed on initially oxidized (pO_2 ~ 3.6 x 10^{-9} Pa) Alloy 800 specimens coated with various deposit mixtures and exposed to gas mixture with pO_2 and pS_2 of ~ 5.4 x 10^{-7} and ~ 1.6 x 10^{-3} Pa, respectively.

Figure 7.27 shows SEM photographs of morphologies of initially preoxidized Alloy 800 specimens coated with deposit *mixtures 1–4* and exposed at 840°C to a gas mixture with oxygen an sulphur partial pressures of ~ 5.4 x 10^{-7} and ~ 1.6 x 10^{-3} Pa, respectively. Preoxidation of the specimens in a low pO_2 environment produced thin, chromium rich oxide scales on the specimens. Subsequent exposure of the specimens to the deposit mixtures in the presence of an SO_2-containing gas mixture had no deleterious effect on the preformed oxide scales, indicating that sulphidation of the Cr_2O_3 scale (once developed) is extremely slow. However, such thin oxide scales may not offer protection in the erosive environment of the FBC systems, especially over tens of thousands of hours of service required of the tube banks.

In the *precarburization* treatment, the gas mixture (5 vol%. CH_4–H_2) established a carbon activity of 1. In such an environment, carburization of the alloy specimens simulates *carbon pick-up* in the alloy because of the deposition of unburnt carbon (i.e. in the vicinity of coal feed ports or due to incomplete combustion) on the alloy components. The precarburization for ~ 64 h at 840 °C resulted in extensive precipitation of (Cr, Fe) carbides in the alloy. As a result, the effective chromium concentration and activity in the alloy decreased and caused the alloy to behave (from a scaling standpoint) as a *medium chromium* alloy. For example, if carburization proceeds via the reaction.

$$\frac{23}{6}C_{r(\gamma)} + C_{(\gamma)} = \frac{1}{6}C_{r23}C_6 \qquad \text{(in mixed carbide)}$$

where Cr(γ) and C(γ) are *concentrations of chromium and carbon in the austenite phase*, the free-energy change for the reaction, as a function of temperature and the equilibrium constant for the reaction

$$K = \frac{a^{\frac{1}{6}}C_{r23}C_6}{aC_{(\gamma)}aC_{r(\gamma)}^{\frac{23}{6}}}$$

can be used to calculate chromium activity/concentration in the austenite phase for a given level of carbide precipitation. At a carbon activity of ~ 0.3, up to which $M_{23}C_6$ carbide is stable, precipitation of carbides can lower the chromium content in the matrix phase to less than 15% wt. At carbon activity above 0.3, M_7C_3 and M_3C carbides will precipitate and further decrease the chromium activity in the austenite phase.

Figure 7.28 shows SEM photographs of initially carburized Alloy 800 specimens that were coated with deposit *mixtures 1–4* and exposed at 840°C to a gas mixture with pO_2 and pS_2 of ~ 5.4x10^{-7} and ~ 1.6x10^{-3} Pa, respectively. Exposure of the *precarburized* specimens led to the formation of non-protective surface scales in

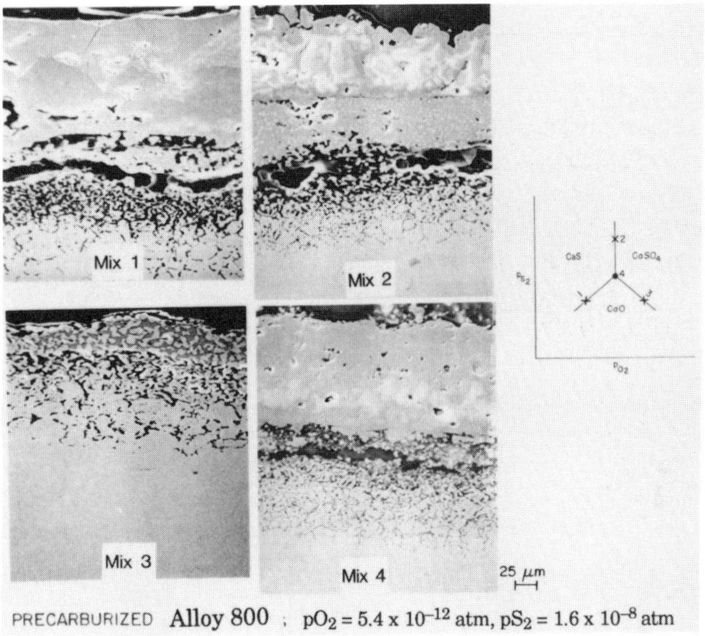

Figure 7.28 Morphologies of scale layers that developed on initially carburized Alloy 800 specimens coated with various deposit mixtures and exposed to gas mixtures with pO_2 and pS_2 of ~ 5.4×10^{-7} and ~ 1.6×10^{-3} Pa, respectively.

the presence of all the deposit mixtures investigated. A comparison of photographs in Figs. 7.26 and 7.28 shows that precarburization has little effect (except in *Mix 3*) on the morphology of the scale layers and that the layers are somewhat thicker in the precarburized specimens. Sulphidation is the mode of attack in this alloy either with or without precarburization treatment. The difference in performance of the alloy in *mix 3* can be attributed to the effective chromium concentration (which will be less in the precarburized specimen than in the untreated specimen) in the austenite matrix. The rôle of deposit in the corrosion performance of materials in the fluidized bed combustion environments can be seen from the data shown in Fig. 7.29 for several commercial alloys exposed with and without sulphated sorbent deposits [34].

7.6.3 Corrosion in the presence of alkali sulphates

Corrosion of gas turbine materials in the presence of liquid sodium sulphate, either by itself or in combination with sodium chloride, has been a problem in gas turbines; this corrosion process has been termed "*hot corrosion*" to differentiate it from gas phase sulphidation attack. Two types of hot corrosion have been

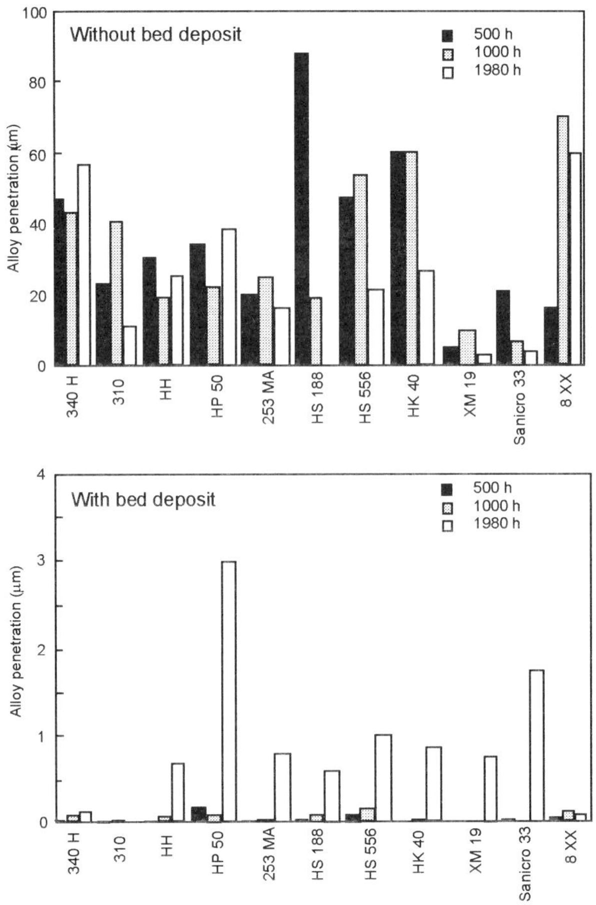

Figure 7.29 Corrosion penetration in alloys exposed in (top) absence and (bottom) presence of sulphated sorbent deposits in FBC environments.

identified: *Type I*, operative at 800°C–950°C, and *Type II*, operative at 550°C–800°C. *Type I hot corrosion can* be split into an *initiation* (or *incubation*) stage and a *propagation* stage. The process, in general, requires the presence of liquid sodium sulphate (melting point 884°C) on the metal surface. In the *initiation* stage, dissolution of the protective oxide scale occurs via a basic fluxing mechanism and the corrosion rates are generally low. In the *propagation* stage, with the protective oxide having been destroyed and not able to reform, the alloy is subjected to sulphidation by inward diffusion of sulphur, leading to accelerated corrosion rates. Extensive reviews of *Type I hot corrosion* have been published in the literature [35,36].

Type II hot corrosion, also known as "*low temperature hot corrosion*", involves eutectics of base-metal sulphates and sodium sulphate and therefore occurs predominantly at lower temperatures, especially in the effluent of the FBC

Table 7.2 Process environments and modes of material degradation

Process/ component	Gas environment	Gas temp. range (°C)	Metal temp range (°C)	Deposit type	Mode of degradation
Pulverised coal boilers	Oxidising	1227–1527	377–727	Alkali sulphates Ash	Alkali corrosion Fouling
Fluidised bed combustion	Oxidising locally reducing	827–1027	377–927	$CaSO_4$, CaO Carbon fly ash	Oxidation Sulphidation Erosion
Gasification	Reducing sulphur	927–1127	377–677	Fly ash Alkalis Chlorides	Sulphidation Fouling
Magneto hydro dynamics	Oxidising reducing	1327–1527	427–677	K_2SO_4 Fly ash/slag	Alkali corrosion Fouling
Gas turbines	Oxidising	577–627	777–927	Alkalis Alkali sulphates Chlorides	Hot corrosion Erosion
Ethylene pyrolysis	Oxidising/ carburising	827–1027	727–927	Carbon Cock	Carburisation Oxidation
Reformer catalyst tubes	Oxidising/ carburising	727–927	727–927	-	Oxidation Carburisation
Nitric acid production	Oxidising	927	827–927	-	Nitridation Hydrogen embrittlement
Refinery plant	Oxidising	727–927	227–477	Sulphur oxides Halides	Oxidation Sulphidation Chlorination Acid dew point
Municipal waste incineration	Oxidising	927–1127	227–527	Chlorides Sulphates Heavy metals	Oxidation Chlorination Stress corrosion
Pulp and paper	Oxidising spent liquor	727–827	527–627	Carbonates Sulphides Sulphates	Molten salt attack
Heat treating	Oxidising carburising nitriding	927–1327	627–927	Salts Carbon gases	Oxidation Metal dusting Thermal fatigue Salt attack
Aluminium remelting	Oxidising	927	527–727	Flue gas Alkalis Halides	Oxidation Chlorination Sulphidation
Fibreglass	Oxidising	827–1027	527–827	Sulphur Alkalis Sulphates	Oxidation Glass attack

environment [37]. For example, the eutectic temperature for sodium sulphate-cobalt sulphate is 565°C. In this case, the transient oxides of cobalt or nickel (which nucleate in the early stage of oxidation in Cr- and Al- containing superalloys) react with sodium sulphate to form eutectic salts that prevent formation of protective chromia or alumina. The corrosion process is strongly dependent on the partial pressure of sulphur trioxide at the melt/scale interface, but the process occurs at much lower temperatures than the melting point of sodium sulphate.

7.7 Service environments of interest

The major modes of materials degradation in the *process*, *chemical*, *power generation*, and *petroleum industries* are wide ranging, and candidate materials must be evaluated in specific environments of interest to allow a viable selection of component materials and to establish performance envelopes for adequate service. The extent to which the materials undergo corrosion is strongly influenced by the prevailing chemistry of exposure environment, temperature and deposits. In addition, large extrapolations of short time data are generally needed to evaluate life assessment of components; this is a concern because of the *breakaway corrosion* that can be triggered in materials for a variety of reasons, e.g. temperature excursions, feedstock variations, malfunctions in functional components. Even though long term predictions on material performance are difficult, a mechanistic understanding of fundamental processes that can occur in a specific environment is a valuable aid in selecting materials. One can identify several processes in which the exposure environments can be characterized and type of deposits can be specified, thereby establishing possible modes of material degradation. Table 7.2 lists exposure environments, deposit types and modes of degradation for a number of chemical, petrochemical, power generation, and process industries of interest.

7.8 Summary

This Chapter has presented a description of many generic areas of material interactions in several process and power generating systems, with examples selected from several to elucidate the corrosion performance of materials. Generally, a marked benefit could be achieved from a fundamental understanding of the underlying material degradation process. This approach provides significant improvements in the technology base for selecting materials for specific applications and for establishing performance envelopes for these materials for long-term service.

References

1. Natesan K, "High-Temperature Corrosion", Proc. Conf. on Prevention of Failures in Coal Conversion Systems, eds. T. R. Shives and W. A. Willard, *U.S. Department of Commerce, National Bureau of Standards Special Technical Publication* 468, p159, 1976.
2. Chopra O K and Natesan K, "Thermodynamic Equilibria of multicomponent Gas Mixtures", *High Temp. Sci.*, **9**, 243, 1977.
3. Pettit F S, Goebel J A, and Goward G W, "Thermodynamic Analysis of the

Simultaneous Attack of Some Metals and Alloys by Two Oxidants", *Corros. Sci.*, **9**, p903, 1969.

4. Natesan K, "Corrosion and Mechanical Behaviour of Materials for Coal Gasification Applications", *Argonne National Laboratory Report ANL*–p80, 1980.
5. Natesan K, "Corrosion of Metals in Coal–Gasification Environments", *Proc. Conf. on Corrosion/Erosion of Coal-Conversion System Materials*, ed. A. V. Levy, National Association of Corrosion Engineers, Houston, p222, 1979.
6. LaBranche M, Garret-Reed A, and Yurek G J, "Early Stages of the Oxidation of Chromium in H_2–H_2O–H_2S Gas Mixtures", *J. Electrochem. Soc.*, **130**, p2805, 1983.
7. Natesan K, "High-Temperature Corrosion in Coal Gasification Systems", *Corrosion*, **41**, p646, 1985.
8. McNallan M J, Oh J M, and Liang W W, "High Temperature Corrosion of Metals in Argon–Oxygen–Chlorine Mixtures", Proc. JIMIS–3 Conf. on High Temperature Corrosion of Metals and Alloys, *Trans. Jpn. Inst. Met. Supplement*, **24**, p363, 1983.
9. Ihara Y, Sakiyama K and Hashimoto K, "The Corrosion Behaviour of Austenitic Stainless Steels in Hydrogen Chloride Gas and Gas Mixtures of Hydrogen Chloride and Oxygen at High Temperatures", *ibid*, p669.
10. Tiearney Jr T C, and Natesan K, "Sulphidation–Oxidation of Advanced Metallic Materials in Simulated Low-Btu Coal Gasifier Environments", *Oxid. Met.*, **17**, p81, 1982.
11. Malik A U and Natesan K, "Oxidation and Sulphidation Behaviour of Fe–20Cr–16Ni–4Al–1Y_2O_3 Oxide–Dispersion–Strengthened Alloy", *Oxid. Met.*, **34**, p497, 1990.
12. Natesan K and Baxter D J, "The Rôle of Zr and Nb in Oxidation/Sulphidation Behaviour of Fe–Cr–Ni Alloys", *Argonne National Laboratory Report ANL/FE*–90/2.
13. Hindham H and Whittle D P, "Microstructure, Adhesion and Growth Kinetics of Protective Scales on Metals and Alloys", *Oxid. Met.*, **18**, p245, 1982.
14. Ramanarayanan T A, Raghavan M and Petrovic-Luton R, "Al_2O_3 Scales on ODS Alloys", Proc. JIMIS–3 Conf. on High Temperature Corrosion of Metals and Alloys, *Trans. Jpn. Met. Supplement*, **24**, p199, 1991.
15. Lessing P A and Gordon R S, "Creep of Polycrystalline Alumina, Pure and Doped with Transition Metal Impurities", *J. Mater. Sci.*, **12**, p2291, 1977.
16. Irving G N, Stringer J and Whittle D P, "The High Temperature Oxidation Resistance of Co–Al Alloys", *Oxid. Met.*, **9**, p427, 1975.
17. Pfeiffer I, "Kristallstruktur und Zusammensetzung von Oxydeschechten auf Eisen–Chrom–Aliminium–Legierungen in Abhangigkeit von der Oxydischichtdicke", *Z. Metallkd.*, **53**, p309, 1962.
18. Pivin J C et al. "Oxidation Mechanism of Fe–Ni–20–25Cr–5Al Alloys – Influence of Small Amounts of Yttrium on Oxidation Kinetics and Oxide

Adherence", *Corros. Sci.*, **20**, 351, 1980.
19. Zelanko P D and Simkovitch G, "High Temperature Sulphidation Behaviour of Iron–Based Alloys in Hydrogen Sulphide–Hydrogen Gas Mixtures", *Oxid. Met.*, **8**, 343, 1974.
20. Huang T T et al "Formation of Aluminium Oxide Scales in Sulphur-containing High Temperature Environments", *Metall. Trans. A*, **16A**, p2051, 1985.
21. Natesan K, *Argonne National Laboratory, unpublished work.*
22. Gesmundo F, and de Asmundis C, "Corrosion of Two Nickel–Silicon Single-phase Alloys in SO_2 at 600–1000°C", Proc. Ninth Int. Congress on Metallic Corrosion, National Research Council, Canada, **2**, p38, 1984.
23. Meier G H and Gulbransen E A, "Corrosion Mechanisms of Coal Combustion Products on Alloys and Coatings", AR & TD Fossil Energy Materials Program Quarterly Progress Report for the Period Ending September 30, 1985, *Oak Ridge National Laboratory Report ORNL/FMP–85/4*, 379, 1985.
24. Baxter D J and Natesan K, "Mechanical Considerations in Degradation of Structural Materials in Aggressive Environments at High Temperatures", *Rev. High-Temp. Mater.*, **5**, p349, 1983.
25. Stott F H, Chong F M F and Stirling C A, "Effectiveness of Preformed Oxides for Protection of Alloys in Sulphidizing Gases at High Temperatures", *Proc. Ninth Int. Congress on Metallic Corrosion*, National Research Council, Canada, **2**, p1, 1984.
26. Dial R E, "Refractories for Coal Gasification and Liquification", *Am. Ceram. Soc. Bull.*, **54**, p640, 1975, .
27. Reid W T, "External Corrosion and Deposits", American Elsevier, New York 1971.
28. Wright I G, Price C W and Herchenroeder R B, "State of the Art and Science Report on Design of Alloys Resistant to High Temperature Corrosion–Erosion in Coal Conversion Environments", *Electric Power Research Institute Report EPRI–FP–557*, 1978.
29. Sondreal E A, Gronhovd G H, Tufte P H and Beckering W, "Ash Deposits and Corrosion Due to Impurities in Combustion Gases", ed. R W Bryers, Hemisphere Publishing Corp., Washington, DC, 85, 1978.
30. Borio R W, Plumley A L and Sylvester W R, *ibid*, p163.
31. Kihara S, Isozaki T and Ohtomo A, "Laboratory Evaluation of Fireside Corrosion of Superheater Tube in Coal–Fired Boiler", *Proceedings JIMIS–3 High Temperature Corrosion*, Tokyo, p655, 1983.
32. Natesan K, "Role of FBC Deposits in the Corrosion of Heat Exchanger Materials", *High. Temp. Technol.*, **4**, p193, 1986.
33. Natesan K, "Laboratory Studies on Corrosion of Materials for Fluidized Bed Combustion Applications", *Argonne National Laboratory report ANL/FE–90/1*, 1990.
34. Natesan K and Podolski W F, "Materials for FBC Cogeneration Systems", *Proc. First Int. Conf. on Heat Resistant Materials*, 23–26 September 1991,

Fontana, WI, eds. K Natesan and D J Tillack, ASM International, p549, 1991.

35. Stringer J, "Annual Review of Material Science", 7, ed. R A Huggins, Annual Reviews Inc., Palo Alto, 477, 1977.
36. Goebel J A, Pettit F S and Goward G W, "Mechanisms for the Hot Corrosion of Nickel–Base Alloys", *Met. Trans.*, **4**, p261, 1973.
37. Haskell R W, Doring H von E, Le Blanc O H and Luthra K L, "A Mechanistic Study of Low–Temperature Corrosion of Materials in the Coal Combustion Environment", *Oak Ridge National Laboratory Final Report ORNL/Sub/84–00224/01*, 1987.

Chapter 8

Surface degradation of aero-gas turbine engines

R G Wing – Chromalloy United Kingdom Ltd (Division of Chromalloy Gas Turbine Corporation, USA)

8.1 Introduction

Gas turbine engines can vary significantly in both size and power output and are used in a diversity of applications; as propulsion units in automotives, trains, boats, hovercraft, for generation of electricity, pumping of natural gas and oil, and to power both civil and military aircraft; it is this last application that forms the basis of this review.

Aero-gas turbine engines are installed in helicopters, fighter and transport military aircraft, small business jets and large passenger carrying civil aircraft; the smaller engines are used in helicopters and these have power outputs in the range ~ 0.37–2.2 MW, whilst the larger civil engines are approaching 100,000 lbs thrust. The rôles of the different aircraft types vary significantly and it is these variations in usage that will affect the extent of degradation of the aero-gas turbine engine.

Helicopters, by design, operate close to the ground and have a short flight cycle of the order of 1–2 hours; due to the versatility of these aircraft they operate in adverse conditions being sometimes very close to the sea or desert environments. Sea water is highly corrosive and sand is hard and erosive and both can be sucked into the operating gas turbine engine (typical helicopter operating environments are shown in Fig. 8.1 and 8.2).

Military fighter aircraft can also be subjected to adverse environments when on low flight paths although this will not be as severe as experienced by helicopters; however, in war situations, the aircraft will fly on maximum power and this in itself can have an adverse affect on engine life due to the increased engine temperatures.

Both military helicopter and fighter aircraft can be stowed on aircraft carriers or at navy airports; even in a non-flying rôle, the aircraft and engines can be subjected to significant corrosive conditions due to the increased salt content of the local atmosphere.

Passenger carrying civil aircraft generally have a much less arduous existence as the flight times are long and aircraft are away from corrosive and erosive

Figure 8.1 A rescue helicopter operating at sea level ingests a high level of sea water into the engines.

Figure 8.2 A military helicopter operating in a desert environment ingests a high level of sand into the engines.

environments; however, short haul "island hopping" operations (flight cycles typically of one hour or less) will experience increased engine corrosion problems due to the increased exposure time to marine environments.

Gas turbine engines installed in flying aircraft can be exposed to very severe environmental conditions and engine life may be controlled by these factors; helicopter engines are typically designed for 5000 hours, military aircraft for 3000 hours and civil aircraft for 20,000 plus hours, based on *cyclic life* of the major rotatives and *creep life* of turbine blading. Many operators of aero-gas turbine engines have to reject components prematurely due to environmental degradation of component surfaces; this can be a significant cost factor to the operator.

8.2 Gas turbine design

The mechanical arrangements of the basic gas turbine engine are essentially simple in that a single shaft, containing compressor and turbine stages, rotates at the centre line of the engine [1]. The compressor is situated at the front of the engine and progressively compresses the incoming air; the compressed air enters into a combustor chamber where it is mixed with fuel and ignited, and the hot gases are then fed into the turbine at the rear of the engine; part of the energy from combustion is used to drive the turbine (and hence the compressor) whilst the remainder is the power for propulsion. This is either exhausted from the rear, or drives a "*power*" turbine stage.

However, designs of the aero-gas turbine engines have evolved; engines can be used as *turbo-jets*, *turbo-props*, *turbo-fans* and *turbo-shafts*. Current engine designs are sophisticated and engines have multi-shafts with different sections of the compressor and turbine running at different speeds to improve engine efficiency; the newest civil engines now employ a front fan stage to optimize efficiency even further. An older single-shaft helicopter engine, the *Rolls-Royce Gnome*, is shown in Fig. 8.3 and a modern large multi-shaft civil engine, the *Rolls-Royce RB211–535* is shown in Fig. 8.4.

In modern gas turbine engines, the incoming air is compressed (in the compressor) to a ratio of up to 20:1 and the compressor exit air temperature can be 500–600°C; after combustion, turbine entry temperatures are of the order of 1200–1400°C and when exhausted the gas temperature is 600–700°C. Engine high pressure rotor speeds are 30,000–40,000 rpm in small engines and 8,000–10,000 rpm in large engines, but in all instances blade peripheral speeds are close to mach 1.

In its own right, the gas turbine engine produces a gas path environment which can severely affect surface integrity of components without the imposed additional effects of adverse operating environments.

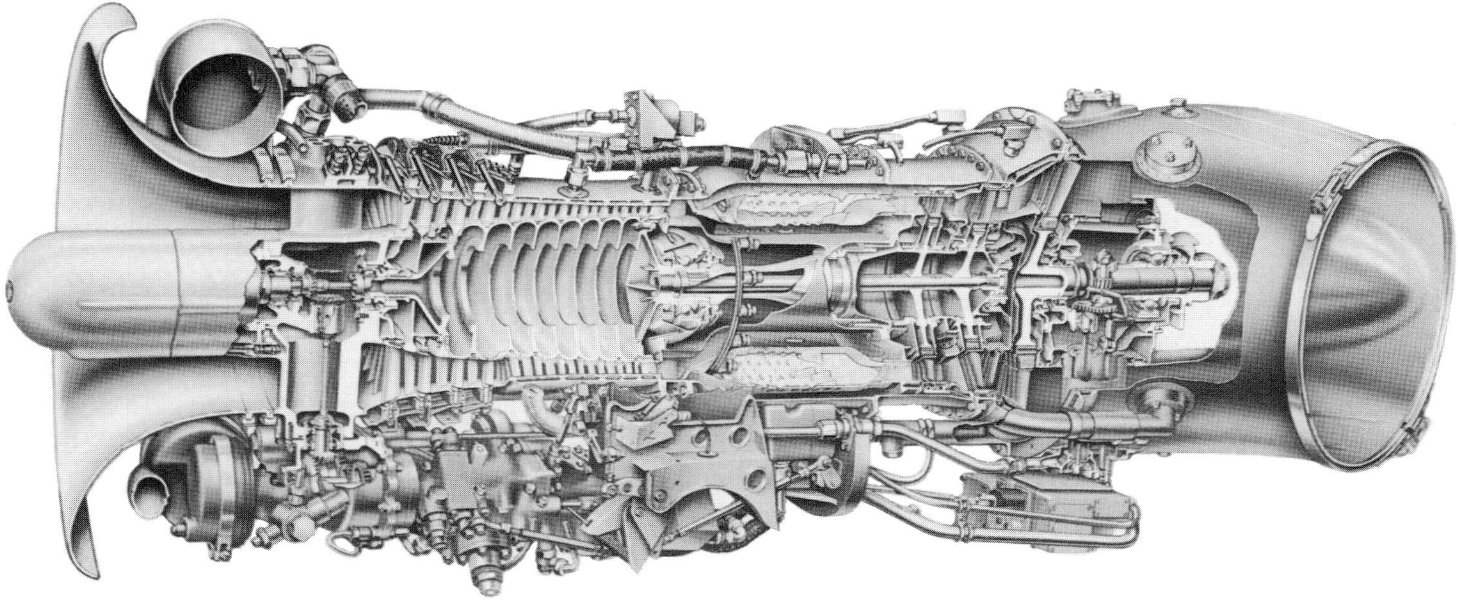

Figure 8.3 Rolls-Royce Gnome helicopter engine.

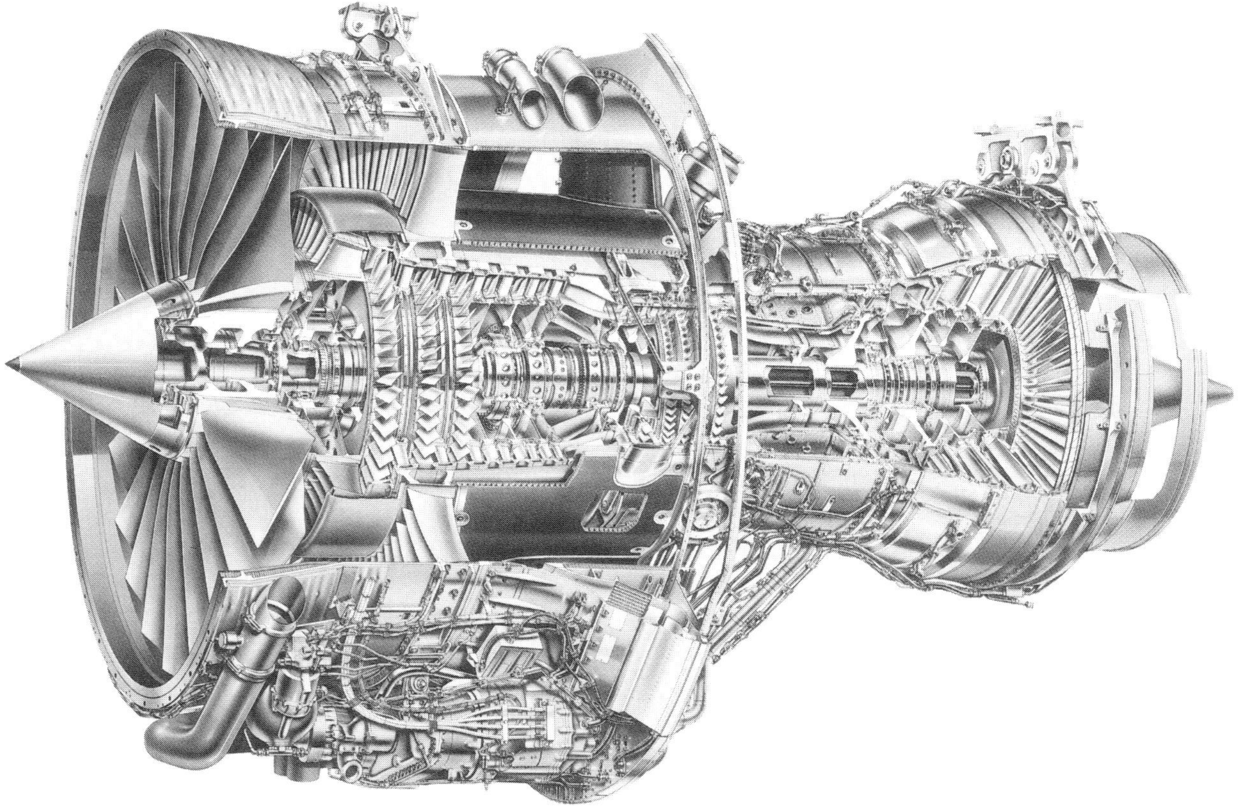

Figure 8.4 Rolls-Royce RB211–535 civil aircraft engine.

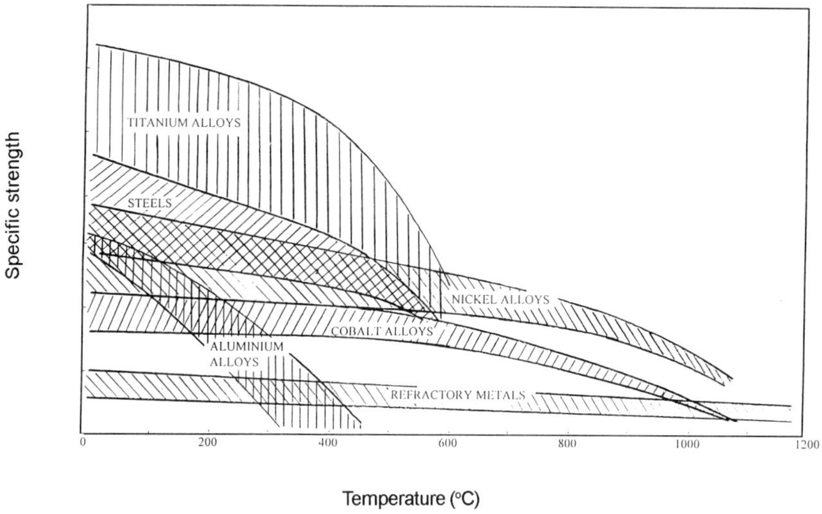

Figure 8.5 Specific strength of various "strong" alloy groups with increasing temperature.

8.3 Materials

Aero-engine materials [2] need to be light and strong and capable of working from ambient up to stoichiometric combustion temperature; in practice, if the specific strengths of various metal alloy groups are examined, then an upper limit of about 1100°C is permissible as a component temperature using proven engineering materials. The generic alloy groups are shown in Fig. 8.5.

Titanium alloys and, to a lesser extent, some steels best meet the requirements for compressor materials; steels are useable up to about 500°C and titanium alloys up to about 600°C. *Aluminium alloys* are also used in compressor applications of some older engines.

Nickel alloys are most suitable materials for combustor and turbine applications, although *cobalt alloys* have been used in some turbine applications where strength requirements are secondary to corrosion resistance; *nozzle guide vanes* in older engines are often made from cobalt alloys.

The newer aero-engines tend to use exclusively titanium alloys in the compressor and nickel superalloys in the combustor and turbine sections; in some "hot" compressors nickel alloys are also selected for the back stages of the compressor. In older engines, prior to suitable titanium alloy manufacturing techniques, steels were generally used for the compressor.

Each of these alloy groups (steels, titanium alloys, nickel superalloys) have been selected primarily based on their *specific mechanical properties* in order to

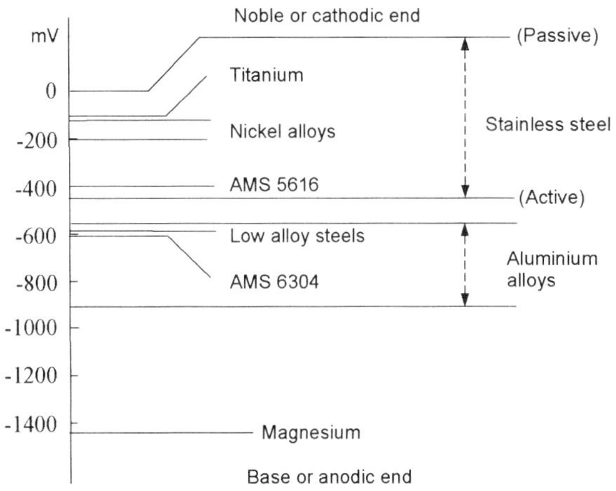

Figure 8.6 Electromotive series relative to a saturated calomel electrode in a 3% sodium chloride solution.

minimise engine weight. However, from the previous considerations of internal engine operating conditions and adverse external environments, these materials also need to resist surface degradation.

The major degradation mechanisms experienced within aero-gas engines are:

- *oxidation* due to the operating temperature of the engine,
- *corrosion* due to ingress of salt water or other contaminants, and
- *erosion* due to contamination of the air flow with hard particles.

There are a multitude of specific and isolated problems that have occurred in aero-gas turbines due to *surface degradation* of components but in this review only the more general examples will be considered.

8.4 Compressor

8.4.1 Corrosion

The majority of older aero-gas turbines have utilized 12–16% Cr martensitic steels for static (stator assemblies) and rotative (blades and discs) compressor components.

The compressor section of the engine can experience severe corrosion conditions due to rain storms, high humidity or close proximity to the sea. These conditions can be both when the engine is flying or when the aircraft is stowed (such as helicopters or flight aircraft on an open carrier deck); more corrosion can be experienced in downtimes if the engines are not thoroughly dried and protected.

It is possible that, due to the many different materials that are built into the gas turbine engine, the corrosion experienced is of a *galvanic* nature. From a consideration of the electro-potential of various alloy systems, it is evident that

Figure 8.7 Pitting corrosion on a 12% chromium martensitic steel compressor blade from a helicopter engine.

steels and aluminium alloys can experience *preferential corrosion attack* (see Fig. 8.6).

The 12–16% chromium martensitic steels, although nominally referred to as being "corrosion resistant", have exhibited severe *pitting corrosion* problems when used for compressor disc and compressor blade applications (an example of compressor blade corrosion is shown in Fig. 8.7 [3]). This attack often covers large areas of the components and is difficult to diagnose due to scale formation;

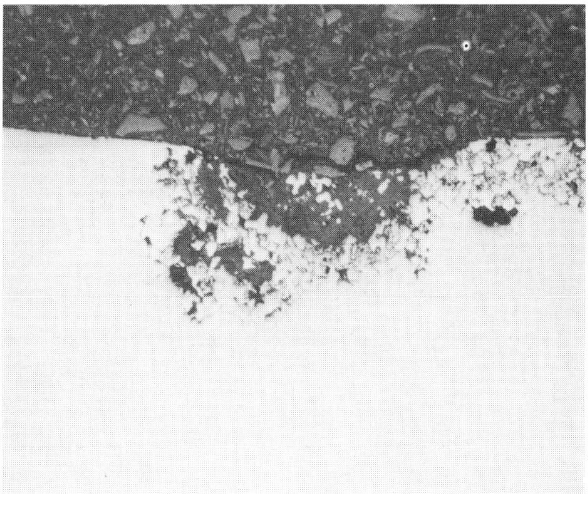

Figure 8.8 Corrosion pitting and associated corrosion damage on a 12% chromium martensitic steel compressor blade.

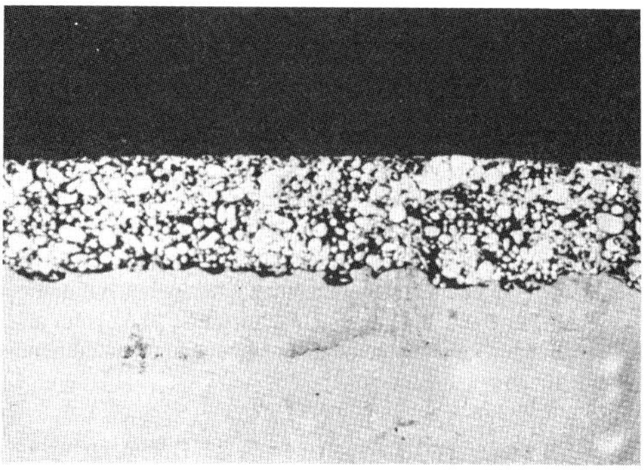

Figure 8.9 Dense, sacrificial aluminium paint system on a 12% chromium martensitic steel compressor blade.

even when components are removed from the engine and the corroded surfaces cleaned, true corrosion damage can be difficult to assess due to localised intergranular attack being located beneath the corrosion *pitting* (typical corrosion pitting and associated corrosion damage is shown in Fig. 8.8). In practice, corroded components are often rejected from further engine usage based on visual assessment using reference pitting standards or known pit depth.

In extreme instances, corrosion pits can act as local stress raisers, and *high cycle fatigue failures* have been experienced on compressor blading due to this reason.

In older engines, corrosion-prone components were normally protected by simple *resin-based paint systems* which sometimes contained aluminium fillers; however,

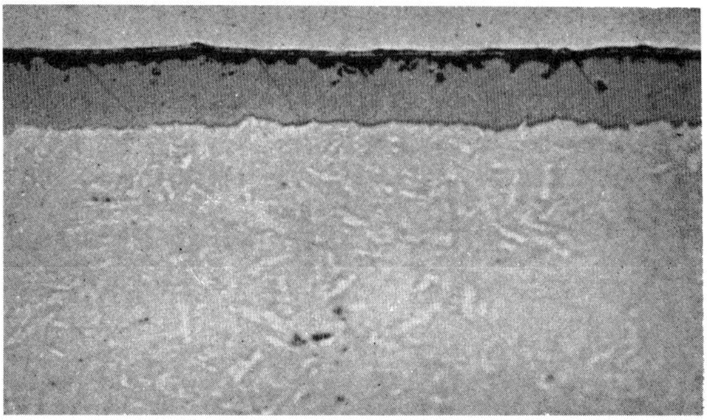

Figure 8.10 Low temperature pack aluminised coating on 12% chromium martensitic steel compressor blade.

these systems were temperature limited, non-conductive and did not provide significant corrosion protection. More recently, inorganic paint systems with aluminium fillers [4] have been developed which are suitable for use up to 500°C and can be made to be *"sacrificial"* to the base material by use of surface compaction techniques (a coating of this type is shown in Fig. 8.9). These types of coating are a significant improvement over the earlier paint systems in that localised unprotected areas, caused by erosion or impact damage, will be sacrificially protected.

Another system used for the protection of steel compressor components is based on the formation of a protective layer at the component surface by diffusion of aluminium in the surface [5]. This is achieved through a low temperature *pack diffusion* process and an *iron aluminide* surface layer of about 20 μm is formed; these coatings normally have an additional 2μm of a phosphate conversion coating which further adds to the corrosion protection and provides a smooth surface finish which benefits compressor efficiency (these features are shown in Fig. 8.10). Coatings of this type are widely used on components which operate in severe marine environments and can be used to about 500°C.

Most modern gas turbine engines have compressors which are predominantly manufactured in *titanium alloys* and these alloys do not experience corrosion problems in a conventional sense; commercial titanium alloys are capable of being used up to about 600°C and, although metallurgically stable at this temperature, surface degradation occurs due to inward diffusion of oxygen. This phenomenon is known as *"α case"* formation and when produced can seriously damage the mechanical properties of the alloy (a typical *"α case"* layer is shown in Fig. 8.11); being a diffusion-controlled process, the depth of the *"α case"* will increase with increasing time and temperature as shown in Fig. 8.12.

Figure 8.11 "α case" formation on the surface of titanium alloy compressor blade.

In addition to the well understood phenomenon of "*α case*" formation, titanium alloys appear to be susceptible to *hot salt stress corrosion cracking* within the temperature range of 250–350°C when exposed to saline environments [6]; many researchers have investigated this phenomenon in the laboratory and have demonstrated significant reduction in mechanical properties of titanium alloys when exposed to salt-containing environments.

Aero-gas turbine engines with titanium compressors, especially those installed in helicopters, have operated in severe marine environments for many thousands of hours and no failures have been reported that can be attributed to the hot salt stress corrosion phenomenon; this is unexpected and poses the question as to whether the dynamic environment present in a running gas turbine engine can be truly represented in a laboratory test.

Laboratory testing indicates that all titanium alloys are prone to this corrosion mechanism, although increasing aluminium content in the alloy will increase sensitivity to the phenomenon and increase of *β-stabilising elements* will decrease it; sufficient engine running has been accumulated with the high aluminium alloys to have expected to see some evidence of the problem.

The mechanism of *hot salt stress corrosion* is complex and is not fully understood, although it is possible that the major alloy weakening is caused by *hydrogen embrittlement*. Some possible explanations as to why titanium compressor components appear not to exhibit the stress corrosion phenomenon are:

- laboratory test conditions cannot be maintained on a component surface due to the engine environment,
- engine running (an isothermal soak period) palliates corrosion damage by allowing hydrogen concentrations to diffuse into the bulk of the material,

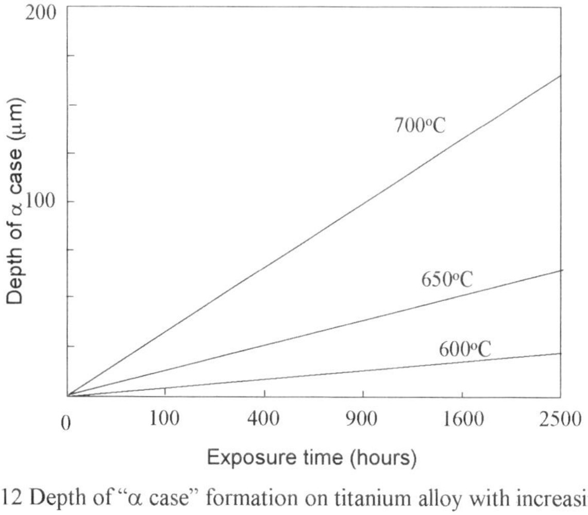

Figure 8.12 Depth of "α case" formation on titanium alloy with increasing time at various temperatures.

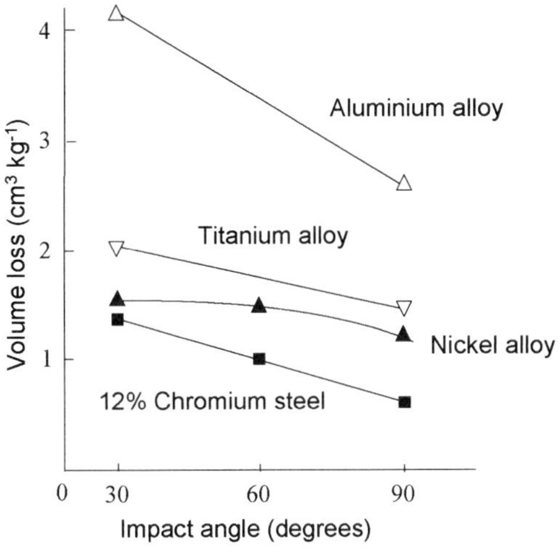

Figure 8.13 Erosion volume loss results of compressor alloys obtained on a laboratory test rig. 40 μm quartz particles at 305 metres/second impact velocity were used to create the erosion condition.

- increasing pressure within the engine suppresses chemical reactions.

The aspect of *hot salt stress corrosion* and "*α case*" formation are relevant to the life expectancy of titanium alloys when used in aero-gas turbine engines; at the present time, "*α case*" formation appears more relevant to engine operation and with a solution to this problem titanium alloys could be used at significantly higher temperatures (this is important as the weight reduction achieved by replacing nickel alloys in hot compressor stages will benefit engine performance). A whole range of coatings have been tested to protect titanium alloys against "*α case*" formation, but to date no coating has been sufficiently successful in preventing oxygen diffusion and maintaining base alloy mechanical properties to be used on production compressor components.

8.4.2 Erosion

In modern aero-gas turbine engines, with almost a total usage of titanium alloys for compressor components, the problems of *aqueous-corrosion* have been eliminated. However, despite the universal drive to increase the use of titanium alloys due to their very good strength-to-weight ratio, these materials are more prone to *erosion damage* than either steel or nickel alloys; the relative erosion resistance of various alloy groups is shown in Fig. 8.13.

Gas turbine engines in most civil and military aircraft operate essentially high above the ground and periods spent close to the ground are only a very small part of the engine life. However, small gas turbine engines in helicopters can operate a large proportion of their life in close-to-ground situations (on troop carrying

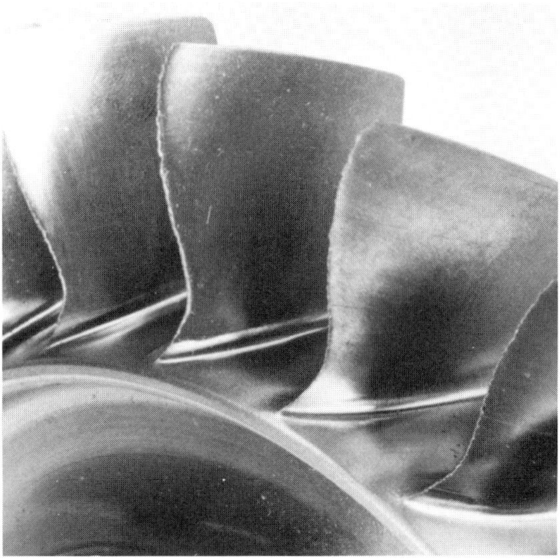

Figure 8.14 Titanium alloy compressor blades exhibiting loss of material at the leading edge due to sand erosion.

missions) and if this is in desert environments then potentially severe erosion damage can occur due to ingress of sand into the compressor; aerofoil damage on a first stage compressor blisc can be seen in Fig. 8.14.

Desert and beach environments will contain a wide variety of different sands, ranging from very fine dust to the extreme of large pebbles. In practice it is the coarse sands that are the most damaging and they contain hard, angular particles across a size range of 0–1000μm.

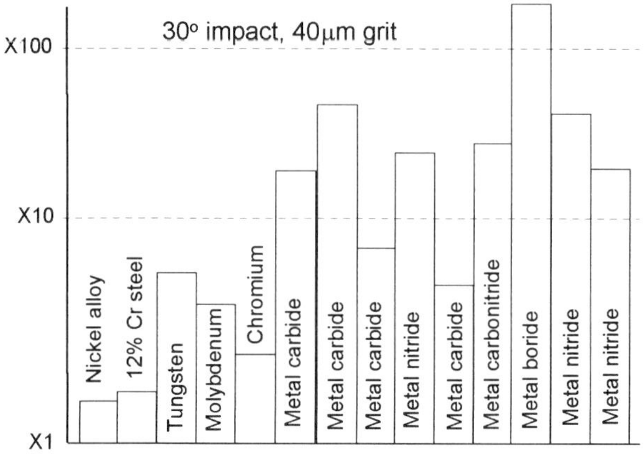

Figure 8.15 Life improvement factor in erosion of various materials and coatings compared to uncoated titanium alloy.

Considerable research effort has taken place to understand the erosion problems experienced by aero-gas turbine engines but often this has been without a total understanding of the effects within the engine. The prime consideration is the level of energy of the impacting particle and how this is dissipated on contact; the view that particle energy can be assessed from air stream velocity is not true as contact with rotating components occurs at significantly higher energies.

When airborne sand particles enter the compressor, the first major erosive condition will be experienced on contact with the first stage of compressor blading; contact will occur between the rotating blade and the incoming particles and at the blade periphery this will be at a speed of about 300 metres per second. This first contact will effectively decrease particle size due to break-up of the larger particles on impact and the mean particle size of the sand aggregate will decrease significantly. With further impacts on successive stages of compressor blading, mean particle size will be further decreased, although to a lesser extent; this means that the *actual erosive environment changes throughout the compressor* and this is an important consideration.

Erosion damage on compressor blading will result in loss of material, especially at the blade leading edges and tips, although the specific pattern of damage will vary between engine types (a typical erosion pattern is shown in Fig. 8.14). As erosion damage increases in the compressor with progressive loss of material from the compressor components, the aero-dynamic efficiency of the compressor will decrease and surging can occur; in some instances, with sufficient loss of material towards the base of compressor aerofoils, the blading can become prone to *high cycle fatigue* and blade failures can occur.

Palliation of compressor erosion damage in helicopter gas turbine engines has received much consideration and various solutions are possible:

- Filters can be fitted to the engine air intake; these systems tend to be heavy and are not generally favoured due to a decreased payload capability of the helicopters.
- *Inlet particle separators* can be manufactured as an integral part of the air intake system. These work on a centrifugal principle in that sand particles are thrown towards the outside of the separator and expelled from the air intake prior to entry into the engine. These systems tend, at their very best, to be only about 90% efficient and ingress of even just 10% of a coarse sand into a compressor can cause significant damage.
- *Base materials* can be selected for the manufacture of compressor components which are inherently more erosion resistant than titanium alloys; both steels and nickel alloys are about twice as erosion resistant, although steels are not favoured due to their reduced corrosion resistance. In practice, some helicopter engines have nickel alloy compressor components in order to maximize both corrosion and erosion resistance despite an increased weight penalty. In the worst erosion conditions, even nickel compressor blading will not survive and compressor lives as low as 10–20 hours have been reported.

- Protective *coatings* do offer an answer to compressor erosion problems if coatings of sufficient resistance could be developed and no adverse effects on base material mechanical properties were associated with their use.

The major aero-engine manufacturers and universities have investigated coating systems for protection of titanium alloys against erosion in considerable depth. High velocity impact rigs have demonstrated that hard coatings, of the order of 50 μm thick, will protect against *small particle impact*; these are typically metal carbides, nitrides and borides (a life improvement factor for various coating groups, compared with an uncoated titanium alloy, is shown in Fig. 8.15). In practical a sense, these systems could be used on the later stages of compressor blading (where the effective particle size has been reduced due to impact with the first stage of rotor blading) providing the coatings do not experience oxidation problems due to the increased temperature towards the back of the compressor.

At the present time, the major problem that needs to be solved is the development of coatings that will withstand *large particle impact*, the condition experienced by the front stage of compressor blading; large particle impact tends to fracture hard coatings within the first few impacts and loss of coating occurs fairly readily. Within an erosive environment produced by multi-size particle sand, the energy of large particles needs to be dissipated without fracture of the coating; in principle, in order to fully resist multi-size particle impacts, then a "*ductile*" hard coating is required.

The most recent work on suitable coatings [7,8] for total erosion protection has concluded that multi-layer coatings consisting of sequential "*hard*" and "*soft*" layers appear most promising. Coatings of this type will resist small particle erosion through the nature of the "*hard*" layers while the energy from the less frequent large particle impacts is dissipated through the "*soft*" layers (a single large particle

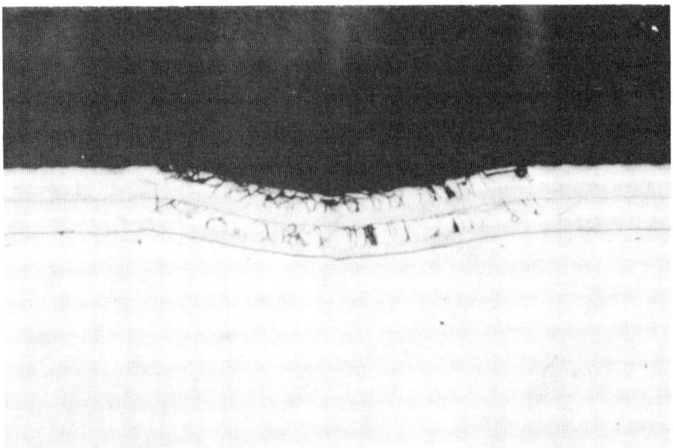

Figure 8.16 Single particle impacted on a multi-layer coating demonstrating how particle energy is dissipated.

impact on a typical coating of this type is shown in Fig. 8.16 – here particle energy is dissipated through plastic deformation and crack formation).

For continuing protection against erosion in an operating compressor, the ideal coating needs to behave in a manner whereby repeated impacts of both small and large particles produce no damage, neither plastic deformation nor crack formation; in essence, the coating will have a "*fatigue limit*" and until this is reached it will remain damage-free.

Some engine trials have been undertaken with titanium nitride coatings on selected stages of compressor blading and these have proven reasonably successful in providing protection against small particle erosion. The more fundamental problem of developing coatings for total erosion protection is still under investigation although, to date, no multi-layer coating systems appear totally effective in combating the worst erosive conditions encountered on the first stage of compressor blading.

The most promising erosion resistant coatings have been deposited using either *chemical vapour deposition* or *physical vapour deposition* methods, but this has highlighted another problem area. Chemical vapour deposition process temperatures are normally above 900°C and these temperatures will damage the mechanical properties of the titanium alloy. New, low temperature processes may have to be developed for the deposition of erosion resistant coatings.

8.5 Combustor

Combustors perform the important task of burning large quantities of fuel with extensive amounts of air and releasing the combusted gases in a smooth and progressive manner suitable for entry into the turbine. The temperature of the combusted gases is of the order of 1800–2000°C and this is too hot for entry into the turbine section; of the air that exists from the rear of the compressor, only about 40% is used in the combustion process with the remainder being used to dilute and cool the combusted gases and to cool the wall of the combustor itself.

Combustor design has evolved with the development of the aero-gas turbine engine with most modern engines having annular combustors; in older engines combustion was achieved through separate combustor chambers placed around the engine circumference. Combustors are normally welded and brazed fabricated assemblies manufactured in nickel superalloy sheet metal materials. These materials need good high temperature strength and resistance to oxidation/ corrosion and thermal cycling.

In most aero-gas turbine engines mean metal temperatures of combustors are fairly low, 800–900°C, due to the effectiveness of the cooling airflow; however *"hot spots"* of 1000–1100°C can exist due to uneven flame patterns and severe localised thermal stress can be produced. For this reason, the primary failure mechanism is normally *deformation* and *thermal fatigue cracking*. Apart from

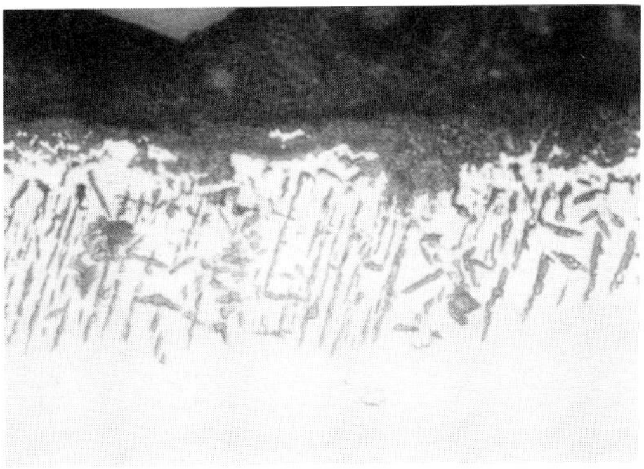

Figure 8.17 High temperature oxidation at the surface of an uncoated superalloy turbine blade.

"hot spot" problems, the materials selected for combustors normally resist the engine environment without a need for coatings; oxidation, sulphidation and carburisation are all surface reactions that have been observed in combustors. The problem of *"hot spot"* damage has been alleviated by the use of *thermal barrier coatings* in selected regions of combustors and life improvements of 3–4 times have been reported. These coatings are heat insulators and applied to depths of about 300–400μm. The first coatings of this type to be used were magnesium zirconate and these were introduced some 20 years ago; more recently, systems based on *yttria partially stabilized zirconia* have been used. (Thermal barrier coatings will be discussed in more detail, when considering the turbine section).

8.6 Turbine

The turbine section of a gas turbine engine consists of various stages of *nozzle guide vanes* (static components that re-direct the hot gas flow) and rotating *turbine blades*; the number of stages of nozzle guide vanes and turbine blades will depend on the particular design of the engine. In earlier engine designs, the various stages of turbine blading were all connected to the same shaft but in more modern engines, there can be *high pressure*, *intermediate pressure* and *low pressure* stages of turbine blade, with each stage rotating at a different speed.

Turbine blades are fixed into turbine discs which act to extract energy from the blades in the form of disc rotation. Turbine discs are outside the gas flow path and for this reason are not subjected to the same degradation processes as turbine blades. In modern gas turbine engines, turbine discs are manufactured in nickel-base superalloy materials and do not experience any major surface degradation

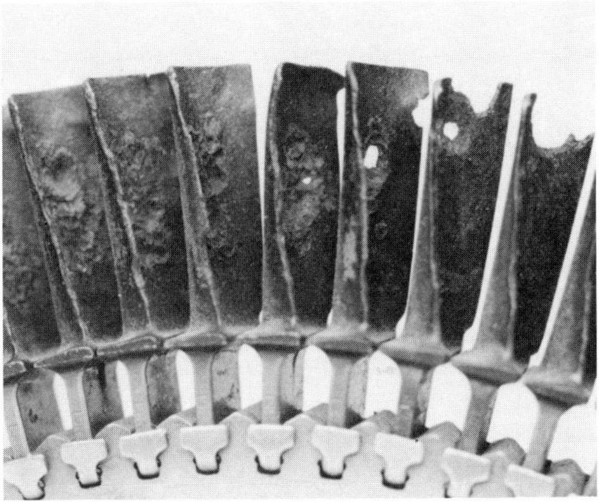

Figure 8.18 Severe hot corrosion attack of long life helicopter engine turbine blades.

during engine operation. However, in older engines some stages of turbine discs were 12% chromium martensitic steels and these can experience pitting corrosion similar to that seen in the compressor; the same protective coating systems are used to extend operational life.

The turbine section is located directly behind the combustor, and hot combustion gases can enter the turbine at temperatures of 1200–1400°C in modern gas turbine engines the combination of hot gases and airborne contaminants – sand and salt

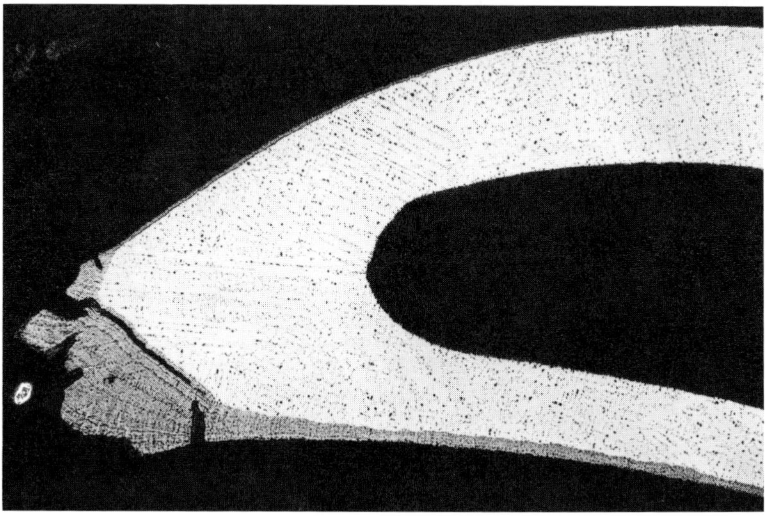

Figure 8.19 Loss of turbine blade leading edge profile due to severe hot corrosion attack.

water from the compressor and carbon from the combustor – plus thermal, centrifugal and cyclic stresses can produce an environment which will severely degrade the surface of turbine components.

The various types of degradation that can occur are:

- *oxidation* (from hot combustion gases),
- *hot corrosion* (*sulphidation*) due to ingress of sea water,
- local *"hot spots"* from an uneven flame pattern in the combustor,
- *erosion* due to ingress of hard particles into the compressor, or
- particulate carbon from the combustion process.

Turbine components in modern aero-gas turbine engines are almost exclusively manufactured in nickel-based superalloy materials, although in some older engines cobalt-base materials are used for some stages of nozzle guide vanes. This selection was possible as stresses are lower in the static components and some strength could be sacrificed for increased corrosion resistance.

Superalloys have been progressively developed over the last 40 years to give improved temperature capability (improved *creep resistance*) for increased performance of the gas turbine engine. However, these developments have resulted from changes in the alloy chemistry (reduction in chromium content) and modern alloys are inherently *less corrosion resistant* than their predecessors.

8.6.1 Oxidation

In terms of the various surface degradation mechanisms that are relevant to turbine

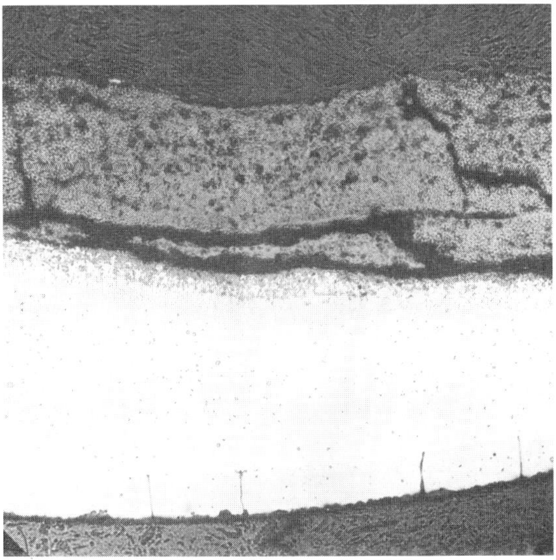

Figure 8.20 Loss of material from the concave surface of a turbine blade due to severe hot corrosion attack.

components [9], *oxidation* is probably the best understood and most predictable. Superalloys have been developed from essentially the Ni–Cr–Al system and need to form stable surface oxide systems; alloy compositions which are essentially alumina-formers will provide much greater oxidation resistance than those which are chromia-formers. Modern superalloys which contain about 5 wt% aluminium for optimum g' precipitate formation need a minimum of 4–5 wt% chromium to ensure alumina scale formation.

Although, in theory, alloy compositions can be designed for maximum oxidation resistance under isothermal conditions, in practice, aero gas turbine engines operate in a cyclic manner; it is under conditions of *thermal cycling* that the normally protective oxides can *spall* resulting in the need then to re-form the protective alumina surface oxide. This process of continually spalling and re-forming the protective oxide can deplete the surface region of aluminium and with time other less protective oxides will form; internal oxidation may occur and oxidation rate will increase (oxidation at the surface of a typical superalloy after thermal cycling is shown in Fig. 8.17).

8.6.2 Hot corrosion

Hot corrosion (or *sulphidation*) [10] is probably the most damaging surface degradation mechanism that can occur in the *turbine section* (*sulphidation*, although often used to describe this corrosion process, is a *misnomer* as it does not really detail the total corrosion process). The hot corrosion process results in a catastrophic form of surface attack that can reduce component section thickness in relatively short exposure times; it can be unpredictable as a corrosion rate due

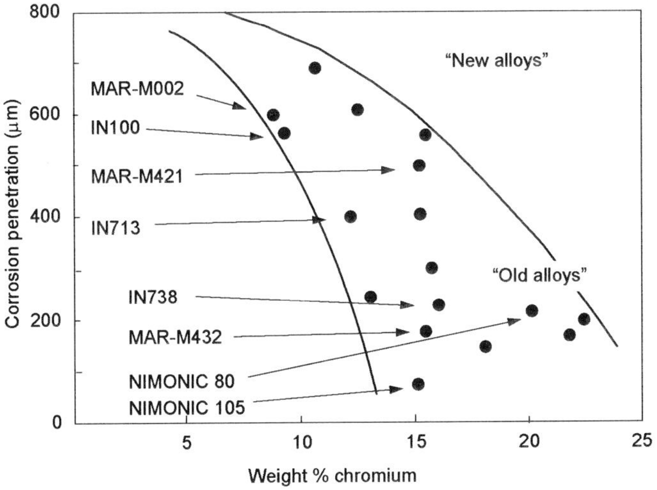

Figure 8.21 Effect of the chromium content of superalloys on the *Type 1* hot corrosion rate.

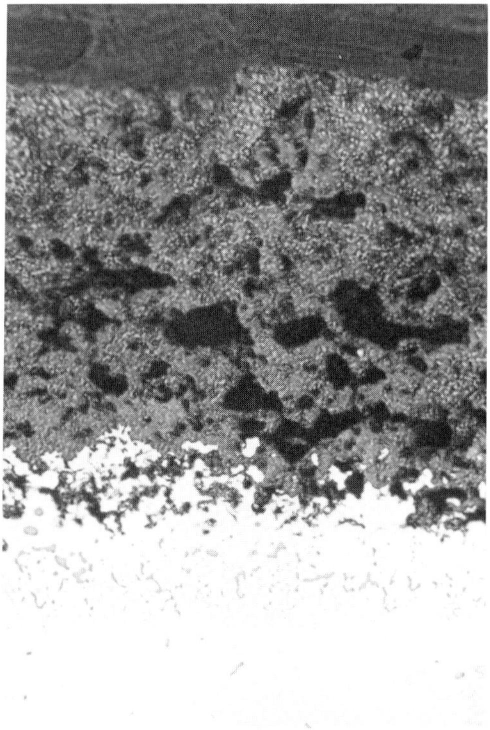

Figure 8.22 Morphology of *Type 1* hot corrosion attack showing thick oxide, denuded alloy zone and metal sulphide layer.

to the continually changing environment that the gas turbine engine will experience in operation. Helicopter engine blades that exhibit severe hot corrosion attack are shown in Fig. 8.18; an example of loss of metal section at an aerofoil leading edge is shown in Fig. 8.19 and an example of loss of section at the mid-chord position is shown in Fig. 8.20.

Hot corrosion results from the deposition of molten sodium sulphate onto the surface of the turbine component. Ingress of salt-laden air into the compressor of the engine from operation in marine environments is the normal source for sodium sulphate as sea water contains about 10%; high sulphur fuels when combusted can also add to the sulphur content, although most commercial aviation fuels are relatively low in sulphur and this is not normally a significant contributory factor in aero-gas turbines.

Hot corrosion has been studied in considerable detail by a great number of researchers and there is still debate on the exact nature of the mechanism of attack. However, the basic reactions within hot corrosion can be seen as being:

- oxygen is removed from the molten sodium sulphate by metal oxide formation leading to metal sulphide formation within the alloy close to the surface,

- removal of sulphur from the sodium sulphate increases the oxide ion concentration in the molten layer which then can flux the normally protective aluminia oxide, and
- as the protective oxide scale is now ineffective, *accelerated oxidation* of the alloy can now take place.

The development of superalloys to progressively increase strength at temperature has been achieved by a re-definition of alloy chemistry; however, one of the more important changes has been the continual reduction in chromium content to achieve this end (earlier superalloys contained 15–20 wt% chromium whilst current alloys contain only 6–8 wt%). This trend in reducing chromium content has resulted in alloys which have much less resistance to hot corrosion attack as shown in Fig. 8.21.

High chromium alloys may better resist hot corrosion because chromia acts as a *"buffering"* agent to the corrosion process, in either consuming oxide ions on the basic side to form chromate, or to contribute oxide ions from chromate to neutralise acidic species. In either instance, a more neutral medium will exist which aids formation of protective oxides.

From observations made on corroded engine components and specific test data obtained from burner rigs, there is a general consensus on the temperature dependence of hot corrosion attack. It is widely accepted that hot corrosion can occur as either one or two corrosion morphologies and these are designated:

Type 1 - corrosion rate peaks at about 900°C
Type 2 - corrosion rate peaks at about 700°C.

Both types of attack are typified by the formation of a thick, loosely adherent oxide scale on the component surface, but the features of extensive alloy denudation and sulphide formation, seen in *Type 1* hot corrosion, are absent in *Type 2*. Although peak corrosion temperatures have been designated for *Type 1* and *Type 2* corrosion, these can vary somewhat, dependent on alloy composition and specific environmental conditions.

In terms of modern aero-gas turbine engines, actual blade metal temperatures are normally too hot for *Type 2* hot corrosion and only *Type 1* is observed. The morphology of *Type 1* hot corrosion is shown in Fig. 8.20. Despite the continuing presence of a corrosive environment, as temperatures rise above 900°C, the corrosion rate will decrease and the hot corrosion mechanism appears ineffective at about 950°C. With a continuing rise in temperature above 950°C an accelerated oxidation behaviour is observed in the presence of sodium sulphate salt, although this is not of the same catastrophic nature as seen with genuine hot corrosion.

8.6.3 Erosion

Erosion of turbine components is not normally a significant problem in modern aero-gas engines, although some instances have been reported. Erodents that can affect the turbine are either contaminants that have entered through the compressor

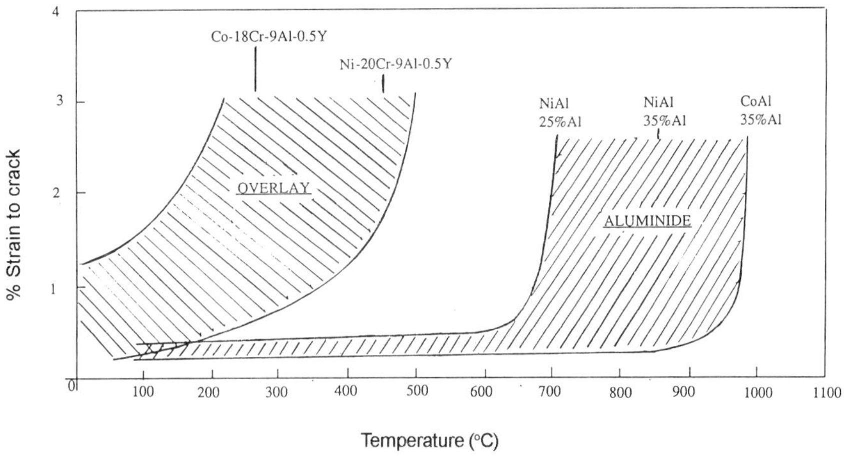

Figure 8.23 Comparative ductility of MCrA1Y overlay and aluminide coatings.

or small carbon particles that have been generated in the combustor. Sand particles that enter the aero-gas turbine engine will degrade the compressor as previously described and it is normally the problem of compressor surging that will cause the engine to be rejected from operation. However, during this period, very fine erodent particles can enter the turbine in one of two ways:

- with the hot gas from the combustor, or
- in the cooling air used to cool the inside of turbine blading.

It is the latter occurrence that is potentially more significant, as the fine dust can progressively *block turbine blade cooling holes* leading to increase in metal temperatures and possible *stress rupture failure*.

Carbon erosion can cause metal loss at the leading edges of turbine blades which results in a loss of aero-dynamic efficiency or in the worst instances causes penetration into the inner cooling passage of the blade. The generation of carbon particles is due to incomplete combustion and, once recognized, can be rectified by design changes to the combustor.

8.7 Turbine coatings

The previously mentioned degradation mechanisms of *oxidation* (including "*hot spot*" damage) and *hot corrosion* experienced on turbine components are countered in the gas turbine engine by providing the alloy surface with a protective coating. These coatings can be divided into three generic types:

- *diffusion coatings* that are formed at the component surface by diffusion of selected elements into the surface,
- *overlay coatings* which are alloys of specific composition developed to resist

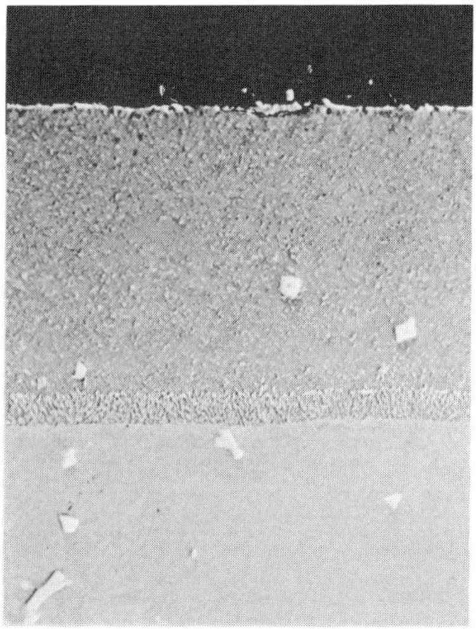

Figure 8.24 Aluminide coating formed on a superalloy material by pack diffusion process.

oxidation and hot corrosion and are applied directly to the component surface,
- *thermal barrier coatings* which are insulating ceramic materials and when applied to a component surface will reduce metal temperatures.

There are many different coatings from the above groups that are used in aero gas turbine engines but all must meet certain criteria, these being:
- to provide good protection against the environment,
- compatability with turbine blade alloy chemistry and have low inter-diffusion rates,
- compatability with the substrate in terms of thermal expansion and are sufficiently ductile at operating temperatures to withstand thermally induced stresses, and
- ease of application and relatively low cost in relationship to life improvement gained.

8.7.1 Diffusion coatings

These coatings [11–13] are produced by diffusing a chosen element into the surface of the component to form a surface layer which is resistant to the gas turbine engine environment.

Aluminising is the most widely used coating of this type and was developed as a pack cementation process some 40 years ago. In this process, aluminium is diffused into the nickel superalloy surface to form a protective *nickel aluminide* layer, normally within a thickness range of 25–75 μm; a similar coating, *cobalt aluminide*, can be produced on cobalt alloys using the same process.

In the pack aluminising process, components are packed in a retort with aluminium, a reactive halide and alumina. The retort is heated in an inert atmosphere within a temperature range of 700–1100°C for a number of hours (exact process conditions will be determined by the pack composition). Aluminium halide gas is produced at the process temperature and this will react with the substrate to form the nickel or cobalt aluminide layer.

Aluminide coatings normally contain aluminium typically within the range 30–35 wt % and are protective due to their ability to continually replenish the aluminia scale that is formed at the surface of the coating. If aluminium content is increased too much the coatings will exhibit increased brittleness and this may increase the tendency for cracking when used in the engine; most aluminide coatings will exhibit ductile behaviour above 700°C, but if increased ductility is required then *MCrAlY overlay coatings* will provide it (see Fig. 8.23).

The main environmental protection used on turbine blades and nozzle guide vanes in aero-gas turbine engines has been aluminide coatings and service history has demonstrated their worth; aluminide coatings with additions of chromium and silicon have also been used for increased *hot corrosion* resistance. However, with the requirements for longer component lives and operation at higher metal temperatures, co-diffusion between the component alloy and the coating has been seen to be a life-limiting factor. Newer coatings have been developed for increased oxide stability and resistance to diffusion, and these are currently used for protection on high pressure turbine blading in modern engines.

The most significant modified aluminide coating to have been developed is *platinum aluminising* and this has been used in production engines for about ten years. Platinum aluminising is normally produced by initially depositing a thin platinum layer (8μm) onto the component surface prior to aluminising; during the aluminising and subsequent heat treatment cycles a complex microstructure of *nickel aluminide* and *platinum aluminide* phases are formed. There are several

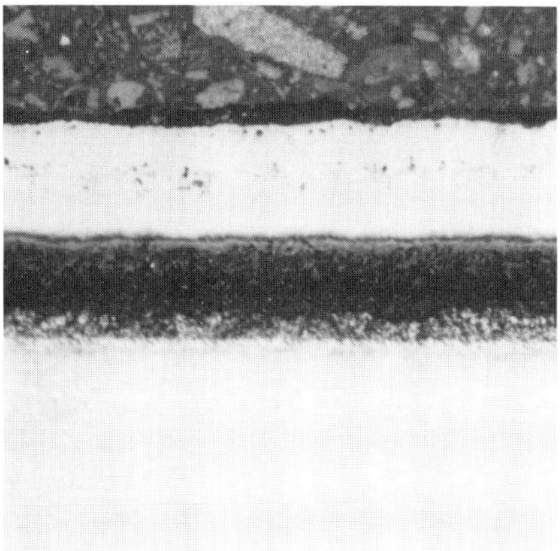

Figure 8.25 Platinum aluminide coating formed on a superalloy material. Platinum enrichment is evident in the outer zone of coating.

variants of platinum aluminising which differ in microstructure and composition, but coatings of this type provide increased resistance to *oxidation* and *hot corrosion*; these improvements can be explained by the reduced co-diffusion characteristics of the coating and the capability to maintain pure alumina at the surface for longer times.

Microstructures of typical aluminide and platinum aluminide coatings are shown in Fig. 8.24 and 8.25; platinum enrichment is evident in the outer layer of the platinum aluminized coating.

There is considerable effort to improve aluminide (and modified aluminide) coatings still further due to their relatively low cost and ease of application; the inclusion of rare earths for increased oxide stability and tantalum as a diffusion barrier is under investigation.

With the introduction of much stricter *environmental regulations* in both Europe and the USA, the disposal of waste pack aluminising powder will become a significant problem. Aluminide coatings are already being deposited using *chemical vapour deposition* methods, but this is on a relatively small scale at the present time; this technique will become increasingly more important as pack cementation processes become outlawed. Chemical vapour deposition uses reactive gases to produce the aluminising environment inside a reduced pressure chamber and waste products are made safe by passing the gases into a scrubber system.

The trends for turbine blades to operate at higher temperatures and for longer periods mean that the *internal cooling* and *film cooling* configurations will become more complex; blades of this design can be coated evenly on the external and internal surfaces by chemical vapour deposition and this will be a further impetus to change away from pack diffusion processes.

8.7.2 Overlay coatings

These types of coatings [14,15] are significant in that they are inherently more ductile than aluminide coatings and are more resistant to co-diffusion with the superalloy; in addition, composition of the protective layer can be designed for specific gas turbine applications. These coatings are based on the *MCrAlY* alloy system, where M is normally either nickel or cobalt or a mixture of both. The actual alloy composition can be optimized for *hot corrosion by increased chromium content* and for *oxidation by increasing aluminium content*; the inclusion of about 0.5 wt % yttrium into the coating is beneficial for improved oxide scale adherence and reduces its tendency to spall during thermal cycling. Aluminium content of *MCrAlY* coatings is normally within the range 6–10 wt % and chromium in the range 20–25 wt %; a coating widely used in aero-gas turbine engines which has both good hot corrosion and oxidation resistance has a nominal composition of Co–32, Ni–21, Cr–21, Al–8, Y0.5 wt %. High aluminium contents will tend to decrease ductility of the coatings and this trend has been shown

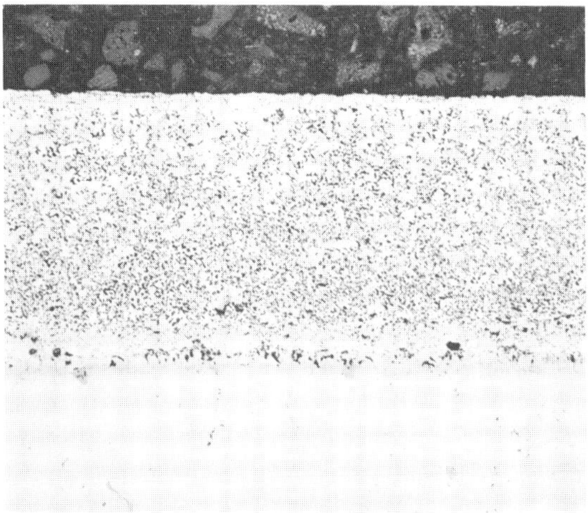

Figure 8.26 MCrA1Y overlay coating deposited on a superalloy material, using electron beam physical vapour deposition.

previously in Fig. 8.23. The microstructure of a typical two phase coating is shown in Fig. 8.26; the duplex nature of the coating is apparent with the major metallurgical phases being *a-cobalt chromium* and *b-cobalt aluminide*.

A large range of *MCrAlY* coatings is already available commercially, but compositional developments are being evaluated for further life improvements and elements such as platinum, tantalum, silicon and hafnium are present in newer coatings.

MCrAlY coatings are normally deposited onto turbine components to a thickness of 100–150 μm, although thicker or thinner coatings may be chosen for specific applications. *MCrAlY* coatings can be used significantly thicker than aluminide coatings due to increased ductility of the system, but extra coating thickness will affect the aerodynamic profile and is *parasitic weight* to the blade and will decrease *creep life*; a compromise between environmental protection, aerodynamic efficiency, and creep life has often to be made. As coatings of this type are applied to the surface as a separate layer, a high temperature diffusion treatment is undertaken to increase bonding to the superalloy; this is normally achieved by heat treating for 2–4 hours within the temperature range 1080–1100°C using a protective atmosphere.

Overlay coatings are normally deposited using either *electron beam physical vapour deposition* (EB–PVD) or *controlled atmosphere/vacuum plasma spraying*; both methods have advantages and disadvantages. Coatings produced by these methods are considerably more expensive than those produced by pack diffusion processes and only used when extreme engine operating conditions are likely to be encountered.

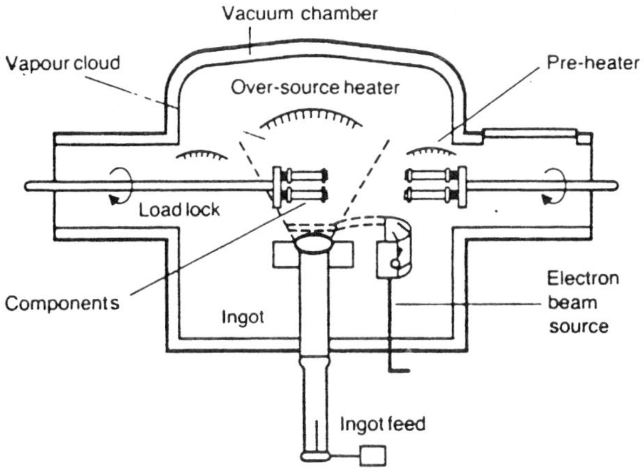

Figure 8.27 Schematic of an electron beam physical vapour deposition coating unit showing the major components.

More recently, deposition techniques such as *composite electroplating*, *sputtering* and *ion-plating* have proved successful in depositing *MCrAlY coatings* but, as of yet, none have moved to production status; further development work may provide alternative production routes.

The *electron beam physical vapour deposition* technique is essentially simple in that coating occurs by positioning the components above a vapour cloud of the desired *MCrAlY* composition inside a vacuum chamber. The vapour cloud is achieved by melting source material at the bottom of the chamber using an electron

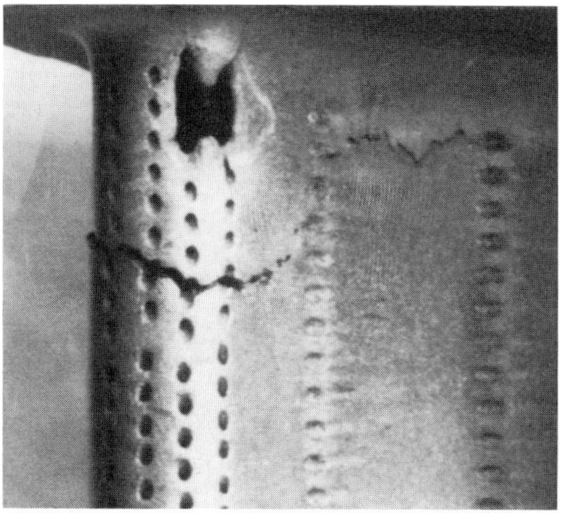

Figure 8.28 "Hot Spot" damage at the leading edge of a turbine aerofoil showing cracking and "holing".

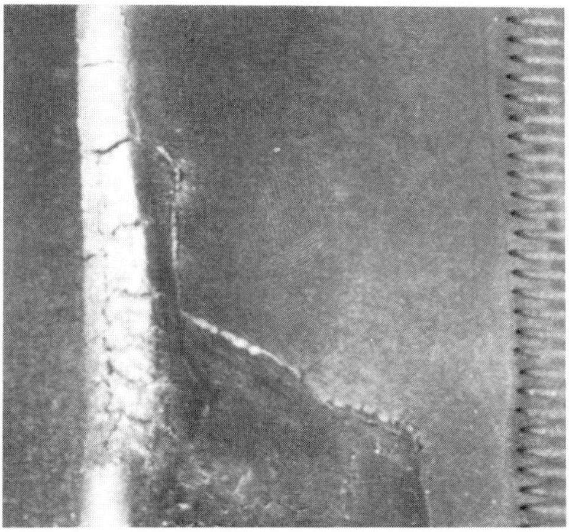

Figure 8.29 "Hot spot" damage at the leading edge of a turbine aerofoil showing cracking and loss of material due to heat erosion.

beam gun; as vapour pressure differs between the various elements, the composition of the source ingot has to be suitably adjusted to take account of this effect. Deposition of the *MCrAlY* alloy occurs by condensation of the vapour on the component surface and coating rates of up to 25 μm per minute are achievable. As the process is predominantly "line-of-sight", the components need to be manipulated in the vapour cloud and this is achieved by rotation and other movement of the holding arm; a schematic of a typical coater is shown in Fig. 8.27.

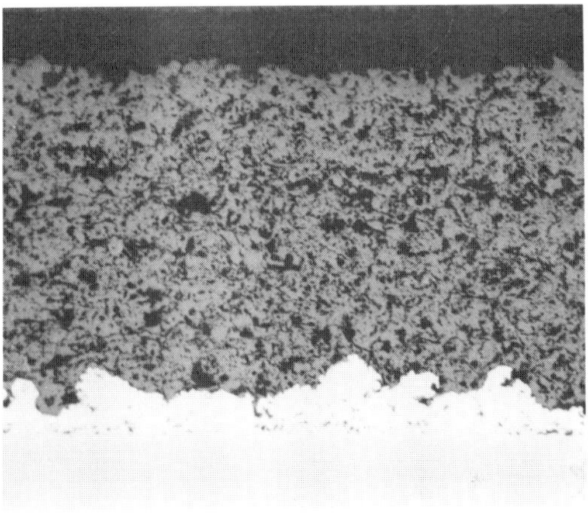

Figure 8.30 Microstructure of thermal barrier coating deposited by air plasma spraying.

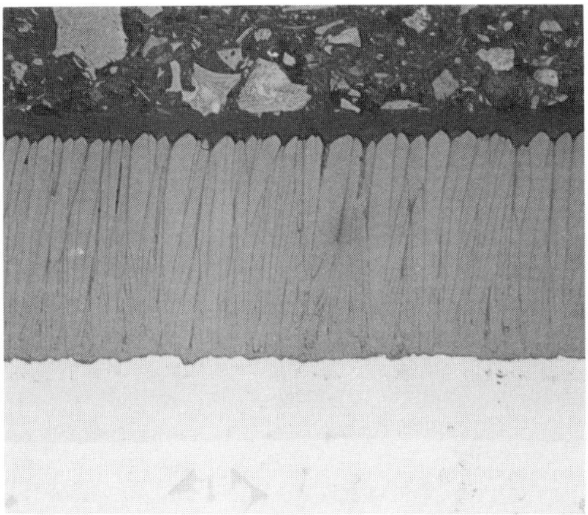

Figure 8.31 Microstructure of thermal barrier coating deposited by electron beam physical vapour deposition.

The total process sequence for coating turbine components is:

- attach components to the fixturing/holding arm in the loading chamber,
- pre-heat components, under vacuum, to a temperature within the range 900–1000°C in the pre-heat chamber,
- coat in the coating chamber for the relevant time to achieve the desired thickness,
- remove from the coating chamber and cool in vacuum.

Following coating, the component surface is glass bead peened and heat treated (within the range 1080–1100°C) to remove the effects of *columnar growth* and to increase *co-diffusion* with the underlying alloy.

Plasma spraying techniques used to deposit *MCrAlY* overlays need to produce coatings that are dense and defect-free for optimum *hot-corrosion*, *oxidation* and *thermal fatigue resistance*. Conventional air plasma spraying methods will deposit coatings that are porous and contain lamellar oxide stringers; the latter can lead to early coating spallation under thermal cycling due to expansion mismatch between oxide and adjacent metal alloy.

Currently, *vacuum plasma spraying* (also called *low pressure plasma spraying*) and *gas shrouded plasma spraying* are used to deposit high quality *MCrAlY* coatings; both techniques will produce dense coatings, essentially free of oxide contamination. *Plasmas* are produced by ionising argon gas (mixed with either hydrogen or helium) inside the plasma gun; *vacuum plasma systems* will achieve a gas velocity of about Mach 2 and *gas shrouded plasma* somewhat less due to air resistance. Metal powder is injected into the plasma where the particles attain a semi-molten state. The plasma gun directs the particles towards the component, impact occurs at high energy, and a dense coating is progressively built-up.

Deposition rates are high and a coating thickness of 100–150 μm can be achieved on a civil engine high pressure turbine blade in 4–6 minutes. Complete coverage of the component is achieved through plasma gun and component manipulation and computer control ensures reproducibility between components.

Vacuum plasma systems have some advantages over *gas shrouded plasma* techniques as pre-heated and transferred arc cleaning are integral with the process cycle. Other plasma systems, such as *high velocity oxy-fuel*, are being developed to deposit high quality *MCrAlY* coatings and these may become production alternatives in the future.

After *plasma spraying*, the coating is normally *shot peened* to ensure complete densification and polished in an abrasive medium to improve surface finish; a heat treatment is also undertaken to increase *co-diffusion* with the component surface.

8.7.3 Thermal barrier coatings

Thermal barrier coatings [16] are ceramic materials with good heat insulating properties. When applied to *turbine blades* and *nozzle guide vanes* these coating systems can even out thermal gradients and reduce metal temperature by 100–150°C. Temperature reduction at the component surface can only be achieved by maintaining a temperature gradient across the coating and this is dependent on the *cooling air flow* inside the component.

The heat insulating properties of thermal barrier coatings can be used to decrease surface degradation of turbine components in one of two ways:

i. Applied locally in "*hot spot*" regions on aerofoils to reduce *thermal strain* and increase resistance to *thermal fatigue cracking* (typical "*hot spot*" damage on blade aerofoils is shown in Fig. 8.28 and 8.29).
ii. Applied as a total aerofoil coating to decrease component metal temperature; this benefit can be used to increase combustion temperature, reduce cooling air flow inside the component or for *creep life* improvements.

The current generation of thermal barrier coatings are based on 6–8% *yttria partially stabilized zirconia* ceramic materials and are deposited to a thickness of 150–300 μm dependent on the particular application. Ceramic coatings have a significant *thermal mismatch* with the underlying superalloy and to compensate for this effect, a *bond coat* is used as an intermediate layer. Bond coats are normally either *MCrAlY* overlays or *platinum aluminising* as both offer good *hot corrosion* and *oxidation resistance*.

Thermal barrier coatings can be deposited by a variety of process methods but the favoured deposition routes for production components are *air plasma spraying* and *electron beam physical vapour deposition*. Each method has its own merits but the significant difference between the two is the morphology of the as-deposited coatings. *Plasma spraying* is a "*layering*" deposition technique and the coatings will exhibit lamellar weaknesses. (Bond coats for air plasma sprayed thermal barrier coatings are usually MCrAlY coatings applied by vacuum plasma spraying although powder particle size is selected to achieve an irregular surface which acts as a mechanical key for the ceramic top coating). Coatings deposited by *electron beam physical vapour deposition* will form by columnar growth with growth direction normal to the component surface; these columns are weakly bonded to the surrounding columns but strongly bonded to the *metallic bond* coat. (Bond coats for electron beam physical vapour deposition thermal barrier coatings are either MCrAlY or platinum aluminide systems as both form a stable alumina surface oxide which acts as the "glue" between the ceramic and the bond coat).

Strain tolerance is an important property of ceramic thermal barriers and has an important rôle in determining the ultimate life of the coating. Plasma sprayed coatings are deposited with a controlled level of porosity for increased strain tolerance. The nature of the columnar growth in the electron beam vapour deposition coatings gives this type very good strain tolerance as the columns can "*open*" and "*shut*" under the thermal strain conditions.

Plasma sprayed and electron beam physical vapour deposition coatings are shown in Fig. 8.30 and 8.31 and the different microstructures are very evident.

The choice of either type of *thermal barrier coating* will depend on its specific application in the aero-gas turbine engine but *electron beam physical vapour deposition coatings* are normally selected for the more exacting aerofoil applications due to certain advantages; these coatings are more strain tolerant, have improved surface finish, retain surface finish longer in engine operation and are more resistant to erosion.

8.8 Future engine developments

With a continuing *increase in combustion temperatures* for better fuel efficiency [17], reduced NO_x emissions and improved performance, the turbine area of the aero-gas turbine engine will become an even more extreme environment, and surface degradation of components will increase. It is unlikely that superalloys can be developed that will have significantly improved *oxidation* and *hot corrosion resistance* over today's materials and, hence, protective coatings will assume an even more important rôle in future engines as gas temperatures increase. Coatings will have to be designed for increased surface oxide stability and reduced co-diffusion characteristics and it will be the addition of specific elements to the general coating composition that will provide these improvements.

The development of *thermal barrier coatings* is one of the most important advances in aero-gas turbine engine technology and temperature improvements already achieved are equivalent to those attained through several generations of superalloys. On very hot turbine components, *thermal barrier coatings* will be used to decrease metal temperature but further improvements to their heat insulation properties and resistance to thermal cycling will be required in future engines. *Thermal barrier coatings will become an integral part of material selection for turbine components* and heat insulating properties will be used to maximum benefit.

Alongside the need for existing turbine materials to operate at higher temperatures in newer engines, there is also the continuing goal of *weight reduction*; new materials such as *ceramics*, *carbon/carbon composites* and *titanium aluminides* may all be used in the turbine area of future engines. The surfaces of these materials are reactive at high temperatures and a new understanding of surface degradation mechanisms will have to be achieved.

Compressor temperatures will also increase as compression ratios increase, but *titanium alloys* will still be the predominant material used in future engine compressors. The surface reactivity of titanium alloys is well understood, but the demands on these materials will increase; compositional changes may decrease surface degradation to some extent but the development of suitable *coatings* is the only real solution. With some types of operation *compressor erosion* will also be a continuing problem and, once again, the development of suitable coatings is a prime requirement.

Future aero-gas turbine engines will experience increased surface degradation of components in the gas path due to increased operating temperatures and the greater use of lighter, more reactive materials. There is a continuing challenge to material scientists to develop alloy systems and protective coatings that will give increased resistance to the gas turbine engine environment; at the present time, future developments in aero-gas turbines are still largely dependent on advances being made in the relevant materials technologies.

References

1. "The jet engine", Technical Publications Dept, Rolls-Royce plc, Derby England.
2. "The development of gas turbine materials", Edited by G W Meetham, Applied Science Publishers Ltd, Barking, England.
3. Camm C, Turbine Support Ltd, Isle of Weight, England, Private Communication.
4. Sermatech International Inc, Ripley, England, commercial brochures.
5. Chromalloy Gas Turbine Corporation, Tilbury, Holland, commercial brochures.
6. Gostellow C R, Weaver M J, NGTE (DRA) Tech. Report R344, 1977.
7. Quesnel E, Paislean Y, Monge-Cadet P and Brun M, "Tungsten and tungsten-carbon PVD multi-layered structures as erosion resistant coatings" *Surf. Coat Tech.* **62**, p474–479, 1993.
8. Darg D, Dimos D, Dyer P N and Stevens R E, "Erosion resistant coatings

containing tungsten carbide by low temperature CVD". Proc. 15th International Conference on Metallurgical Coatings, San Diego, U S A, April 1988.

9. Whittle D P, "High temperature oxidation of superalloys", High Temperature Alloys for Gas Turbines, edited by Coutsouradis, Felix, Fischmeister, Habraken, Lindblom, Speidel. Applied Science Publishers Ltd, Barking, England.
10. Stringer J, "Hot corrosion of high temperature alloys"Ann, *Rev. Mat. Sci.* **7**, p477–509, 1977.
11. Gupta B, Sarknel A K and Seigle L L, "On the kinetics of pack aluminisation", *Thin Solid Films*, **39**, p313–320, 1976.
12. Wing R G, Chadran R, "Chemical vapour deposition coating of gas turbine components", Proc. Surfair X Conf., Cannes, France, 1994.
13. Tawancey H M, Abbas N H and Rhys-Jones T N, "Role of platinum in aluminide coatings", *Surf. Coat. Tech.*, **49**, p1–7, 1991.
14. Lammerman H and Feuerstein A, "PVD overlay coatings for blades and vanes of advanced aircraft engines", Proc. International Gas Turbine Congress, Yokohana, Japan, p269–281, 1991.
15. Fairbanks J W, ASME 76–GT–286, 1976.
16. Sheffler K D and Gupta D K, "Current status and future trends in turbine application of thermal barrier coatings", ASME 88–GT–286,1988.
17. Rickerby D and Low H, "Towards designer surfaces in aero-gas turbines" Proc. European Propulsion Forum, p5.12.1–5.12.12, 1993.

Chapter 9

Fretting and fretting fatigue, incidence and alleviation

D E Taylor – University of Sunderland

9.1 Introduction

Fretting occurs when two surfaces in contact under load and nominally at rest with respect to each other are caused to suffer slight periodic relative motion or slip. The slip is often of very small amplitude, typically less than 20 μm, such as that which might be caused by transmitted vibration. Where the interaction leads to surface damage and wear products determined by chemical reactions with the environment the phenomena is usually termed "*fretting corrosion*".

One of the particularly important and damaging features of fretting, especially where the movement is caused by cyclic stressing of one of the contacting surfaces, is the initiation of *fatigue cracks*. Such "*fretting fatigue*" can lead to significant and marked reduction in the service life of components.

9.2 Fretting and other forms of wear

The question is often asked "*what distinguishes fretting from other types of wear?*" The circumstances in which fretting occurs differ in two important respects from the conditions experienced in other forms of wear [1]. Firstly, *the relative velocity of the two surfaces is usually very much lower* for example, assuming simple harmonic motion with an amplitude of 2.5 μm and a frequency of 50 Hz, the maximum velocity is 0.75 $mm.s^{-1}$ and the average velocity 0.25 $mm.s^{-1}$. Fretting damage has been claimed with a mean relative velocity of only 150 $mm.year^{-1}$. Secondly, because of the very small degree of movement, the surfaces do not come apart as in most other sliding and wear processes and the debris formed as a result of the action (often hard abrasive oxide particles) given little opportunity to escape is trapped in situ and may participate in the causing of further damage. If the amplitude of motion is increased and the frequency remains the same, the velocity will increase and in some situations parts of the contacting surfaces will become exposed the process then being similar to normal wear.

may cause seizure or if the debris can escape it results in loss of fit between the surfaces and a reduction in the clamping pressure which in turn may lead to greater vibration.

In machinery, escaping debris may fall or be carried on to other moving parts and so indirectly cause further wear problems. Less familiar fields which have been struck by the malaise of fretting include steel ropes, surgical implants and electrical components such as switch gear and relays.

The transport industry produces a multitude of fretting problems. With transport problems the vibration arises obviously from the engine propelling the vehicle. Perhaps less obviously the vibration can arise from the road surface in the case of wheeled vehicles or from the passage of the vehicle through the surrounding medium in the case of an aeroplane or ship.

In addition to the fretting damage occurring on the vehicle itself there is the possibility that goods being transported may also suffer. One of the earliest examples of this type of damage was the so-called "*false-brinelling*" in the bearings of cars being transported by road from Detroit to the west coast of the United States. The damage was caused by fretting due to vibration of the stationary vehicle during transportation. Damage of this type can often be guarded against by *filling the bearings with a special lubricant grease* which can be washed out on delivery to be replaced by the normal bearing lubricant. The main requirement for an anti-fretting grease is that it should be able to flow into the contact region and stay there; this ability is governed by its initial consistency and its consistency under shear.

A further example occurred in the transport by rail of "*logs*" of aluminium. These were continuously cast ingots 250 mm in diameter and cut into 3 m lengths, secured in bundles with steel tape and transported in open trucks. On arrival areas of black powder which were originally thought to be graphite were found on the surface of the logs. On subsequent hot extrusion, blistering and skinning occurred on the face of the extrusion. This was due to the black powder which was subsequently diagnosed as *fretting debris*. The remedy in this case was simply to separate the logs with small wooden chucks during transit.

9.5 Press fits

One of the most frequent sites of fretting is in a *press-fit* where a component is shrunk on to a loaded rotating shaft. In the transport industry this is seen in shafts carrying hubs or bearing housings e.g. in railway stock or the fretting of a propeller on the drive shaft of a ship. As the shaft rotates slight movement occurs between it and the press-fitted member resulting in fretting damage near the outer edges of the contact. The shaft is undergoing cyclic-stressing and the dangerous action of fretting in this case is the initiation of a *fatigue crack*. Over 90% of failures in freight car axles are said to be initiated in the wheel seat which is often subjected to high stress concentration and fretting.

Good design is the most important requirement in press-fitted assemblies but since fatigue cracks are initiated in the surface of the shaft further improvement is possible by applying surface treatments. Surface *cold rolling* and *shot peening* have improved performance and other treatments which have been applied are phosphating and metal plating, e.g. chromium. In terms of design the guideline principle in avoiding press-fit *fretting fatigue* problems should be to reduce stress concentrations in the vicinity of the contact. The classic example is that of the wheel-on-axle assembly where the diameter of the wheel seat should be greater than that of the axle with a generous, fillet radius between the two. If this is not possible then a stress relieving groove adjacent to the hub achieves the same objective (Fig. 9.4)[5].

9.6 Riveted joints

A very common form of fretting damage occurs in *riveted, bolted or pinned joints*. When such a joint is vibrated or subjected to reverse bending local movement may occur:

- between the head of the rivet or bolt and the sheet;
- between the two sheets; or
- between the shank of the rivet or bolt and the holes in the sheet.

The main danger in this situation is the initiation of *fatigue cracks*. Such cracks are seen in Fig. 9.5 [2]; originating from fretting damage around a rivet hole. That the cracks are not coming from the edge of the hole can be clearly seen, the fatigue cracks have been more influenced by the fretting than by the stress concentration associated with the hole itself.

Methods to improve the fatigue strengths of such joints include application of suitable *lubricants* or *jointing compounds* and the use of *interference fits* between the pin and the hole.

A situation where fretting frequently leads to failure under conditions of fatigue

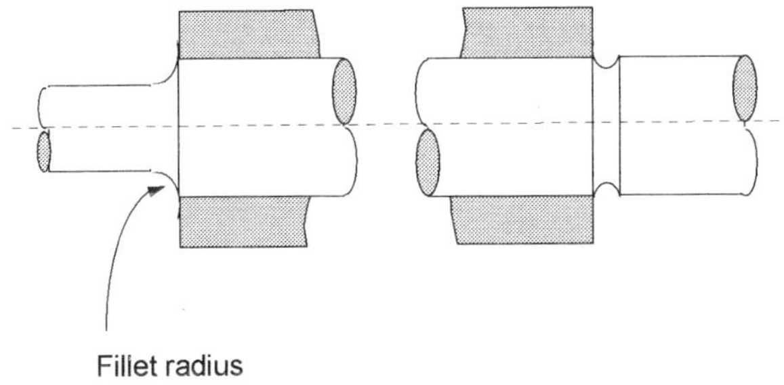

Figure 9.4 Design of press fit to reduce stress concentration.

Figure 9.5 Fatigue cracks propagating from *fretting* damage under rivet head.

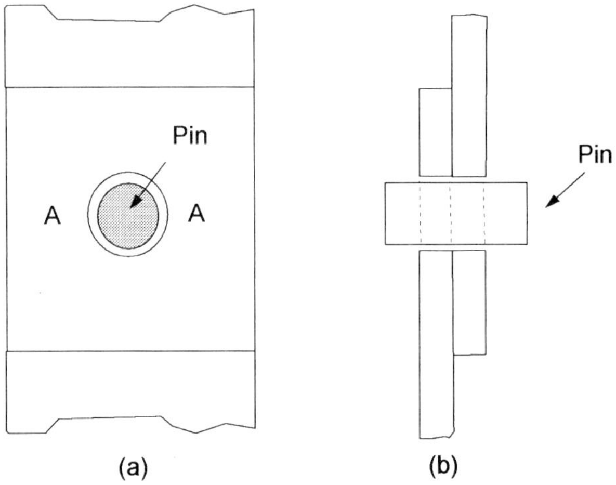

Figure 9.6 (a) Front elevation of pinned joint (b) side elevation.

Figure 9.7 Scavenge pump piston failed by fretting fatigue at centre.

is in the *pin-loaded point* (Fig. 9.6). When such a point is tested in fluctuating tension, fretting occurs at the diametrically opposite sides of the hole, A-A, and a fatigue crack is initiated which propagates through the reduced section to failure. The *machining of flats* on the sides of the pin can avoid contact at this critical maximum stress area [6].

9.7 Clamped and flanged assemblies

Engineering parts which are particularly prone to *fretting fatigue* are the male members of push or interference fits, and members of clamped assemblies that are subjected to cyclic stress. A classic example involved the *scavenge pump pistons* of a large slow speed marine diesel. These pistons were about 1700 mm diameter x 40 mm thick and were machined from steel plate. They were bolted between flanges of upper and lower piston rods. After about six years in service a portion broke from the scavenge pump piston of an engine. The damaged piston is shown in Fig. 9.7. The surfaces of the fracture showed clear evidence of *fatigue crack* propagation, with origins in the upper surface near the outer edge of contact with the flange of the piston rod. It is interesting to note that the cracks originated in the surface between the bolt holes. Enquiries revealed that a number of other pistons had been withdrawn from service for similar cracks. Further failures were avoided by increasing the diameter of the flange and increasing the thickness at the centre of the piston. These measures reduced the cyclic bending stress at the centre of the piston to a level at which fretting cracks would not propagate.

Fretting scars often appear at mating surfaces where intimate metal-to-metal contact is made. Thus any lack of flatness of component parts which may result from manufacture or assembly is likely to influence the position of these scars, particularly so with thin-sheet construction where inadvertent distortion may be

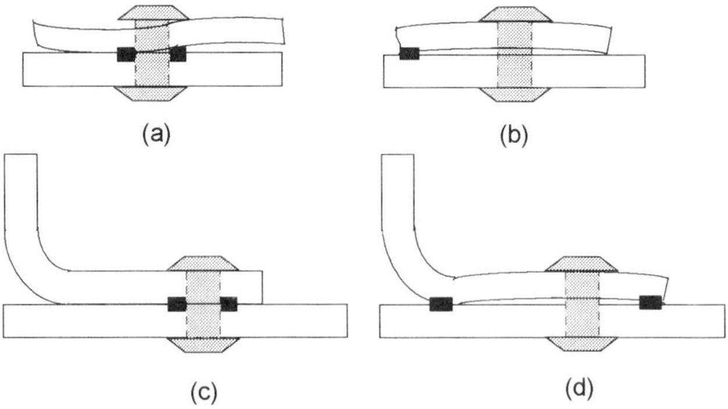

Figure 9.8 Schematics of thin sheet joints illustrating commonly occuring fretting points – danger points indicated.

introduced in the cutting and forming operations. Fastening closure itself can cause appreciable *dishing* around the holes that may lead to non-uniform contact [6]. These types of situations are shown in Fig. 9.8. Other features that present fretting problems are associated with *washers* or *fastener heads* which may not deal uniformly due to distortion in assembly or incorrect geometry and the fairly common circumstance where misalignment or side loads are applied in *fork fitting* (Fig. 9.9)[8].

9.8 Sealing faces

A problem arose fairly recently when fretting was found to be causing loss of fit between the discs in a huge compressor providing air for a wind tunnel. *Fretting wear* had occurred on the relatively small annular surfaces of the 9 feet diameter discs. By *hard metal spraying* it was possible to make good the material lost from the surfaces and to reduce the risk of wear damage in the future.

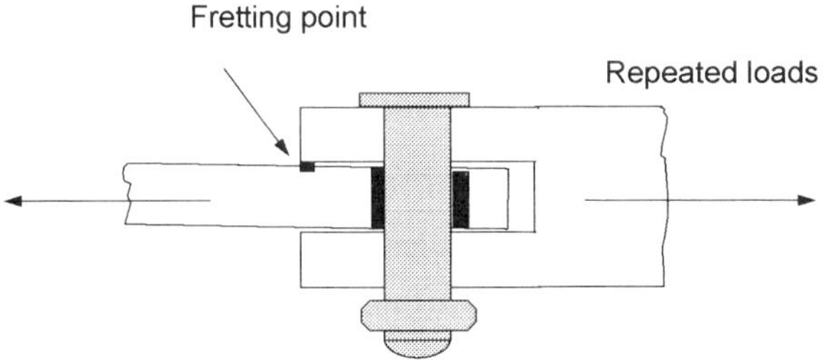

Figure 9.9 Schematic illustrating how misalignment of a fork end fitting can cause fretting fatigue problems – danger points indicated.

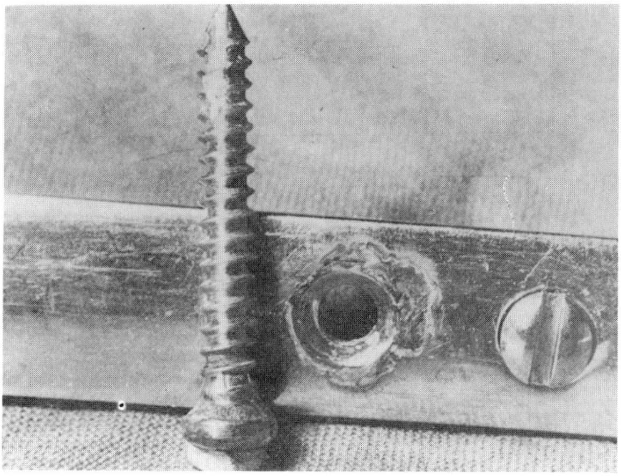

Figure 9.10 Plate and screw removed from septic wound after six months in body.

One of the processes which occurs in the early stages of fretting damage is the formation of *welds* between the two surfaces which in some cases, leads to considerable *adhesion* being developed. A recognised method of minimising fretting damage has been the hardening of the contacting surfaces by *nitriding* or *carburising* in the case of steel components, or coating the surfaces with a hard coating such as *chromium plating* or non-metallic coatings such as *phosphate or anodized coatings*. The function of these coatings is to reduce adhesion between the surfaces and to provide *abrasion resistance* if damage should be inadvertently initiated. Where the surface is being fatigued, the coating may provide additional protection if it contains compressive stresses since these improve the fatigue performance of the component. In recent years the application of hard refractory coatings by flame spraying has developed as an anti-abrasion measure in such components as rocket exhausts. These coatings could be of great value in combating *fretting corrosion* and *fretting fatigue*, particularly in large installations since the coating can be easily applied in situ with portable equipment.

In the case of the compressor discs *molybdenum* was applied to the surfaces using a metal spray gun and not only was the wear reduced but the resistance to *fretting fatigue* was considerably enhanced [7].

9.9 Wire ropes and cables

In construction, ropes consist of strands of wire wound together in a variety of patterns. When the rope is repeatedly flexed or put under tension movement occurs between the contacting strands and this can lead to *fretting corrosion* and when a cable runs over pulleys or fairleads both the sliding and normal loads are increased and the to and fro motion constitutes a *fretting fatigue* situation. Fretting can occur between the wires of a locked-coil type rope, which is of very compact construction having specifically shaped strands which fit together closely and are often used in overhead cable ropeways. If *fretting debris* forms inside the rope it forces the strands apart allowing ingress of the atmosphere or environment and promotes rapid corrosion damage. A way of guarding against *fretting corrosion* within a rope is by suitable choice of a lubricant with which the rope is filled during manufacture.

A characteristic of *fretting fatigue* of ropes and cables is that the first failure frequently occurs on those wires submerged near the centre and trouble may not be apparent until broken ends begin to protrude. This makes effective inspection virtually impossible and consequently control ropes and cables should be *replaced* frequently, irrespective of superficial conditions[8].

9.10 Surgical implants

Metal plates and screws are now frequently used in the human body to aid repair of broken bones. Metal parts are also used as prostheses for example for the hip joint replacement. In most of these cases the metal component is attached to healthy bone by use of screws, the underside of the head of the screw and the surface of the plate or the countersink of the screw hole being in contact. Most repairs are performed on limbs which can undergo a large number of stress reversals

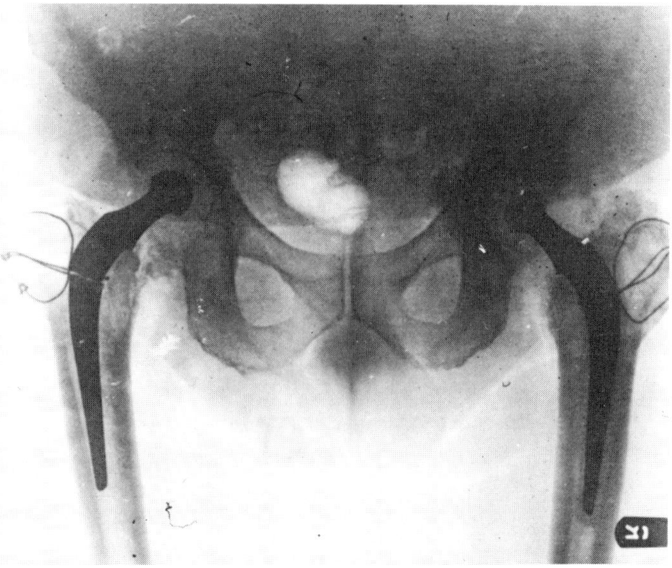

Figure 9.11 Double hip replacement.

in the course of one day, 5000 is quoted as a reasonable estimate. The conditions are therefore those in which *fretting corrosion* can be expected. Fig. 9.10 shows a plate and screw removed from a septic wound after six months in the body. Corrosion within the human body raises problems that are not met elsewhere. Not only is the reduction in strength caused by corrosion important, but even more serious is the effect of the corrosion products on the surrounding tissue, since many base metal ions are poisonous and can destroy living tissue.

Returning to the case of the hip prosthesis, Fig. 9.11 shows an X-ray photograph of a patient who has had a double hip replacement. As part of the operation it was necessary to remove a section of femural bone to allow the implant to be inserted. This bone was later replaced and held in position by stainless steel wires which can be seen in the photograph. After approximately two years the patient complained of severe pain. The implant was removed and it was found that the constant rubbing of these wires on the implant had caused *fatigue cracks* to generate leading to fracture of the prosthesis.

9.11 Conclusion

The foregoing comments indicate that in attempting to alleviate or prevent fretting from reducing the service performance of components or assemblies there are three main approaches; *improved design*, *use of surface treatments or coatings*, and *lubrication*.

An ideal remedy for fretting is to eliminate the vibration which is the cause of the relative motion or slip between the mating surfaces. This can be achieved at the design stage with the ultimate being to eliminate the interfaces by using one solid piece, or more usually by employing a welded structure. In general, however, contacting surfaces cannot be avoided and in *fretting fatigue* situations, where the significance lies not so much in the quantity of the damage produced but in its ability to initiate surface cracks, the aim should be to reduce *stress concentrations* in the contact vicinity.

A wide range of surface modifications has been used as palliatives in *fretting corrosion* and *fretting fatigue* situations. Surface treatments can be applied coatings which include electrodeposited or sprayed metals, anodized or phosphate coatings or be the result of diffusion such as chromising, carburising and nitriding. The objective is usually to increase the hardness and affect the chemical nature of the surface. Several of the above surface modifications and other mechanical treatments such as *rolling*, *shot peening* and *vapour blasting* also induce a *compressive stress*, an important beneficial factor where fatigue is concerned.

The ability of *high viscosity shear resistant greases* to be retained within the contact area allows them to be effective in delaying the onset of fretting damage [9] and can provide a remedy if replenishment is possible at regular intervals. The bonding of solid lubricants such as MoS_2 to metal surfaces can also have ameliorative effects.

The incidences which give rise to fretting problems and the means of alleviation exemplified here are by no means exhaustive. Other fretting situations are legion and for further and more detailed discussion readers are directed particularly to references [1], [5] and [6].

References

1. Waterhouse R B, Fretting Corrosion, Pergamon, Oxford, 1972.
2. Miller W, *Schweiz. Arch. angew Wiss. Techn.* **5**, p300, 1939.
3. Waterhouse R B, Introduction to Fatigue, Fretting Fatigue, Ed. Waterhouse R B, Applied Science, London, 1981.
4. Waterhouse R B, Theories of Fretting Processes, Fretting Fatigue, Ed. Waterhouse R B, Applied Science, London, 1981.
5. Taylor D E and Waterhouse R B, Wear, Fretting and Fretting Fatigue, Metal Behaviour and Surface Engineering, Ed. Curioni, Waterhouse, Kirk, IITT-International, France , 1989.
6. Waterhouse R B, Avoidance of Fretting Fatigue Failures, Fretting Fatigue, Ed. Waterhouse R B, Applied Science, London, 1981.
7. Taylor D E and Waterhouse R B, Sprayed Molybdenum Coatings as a Protection Against Fretting Fatigue, *Wear*, **20**, p401, 1972.
8. Forsyth P J E, Occurrence of Fretting Fatigue Failures in Practice, Fretting Fatigue, Ed. Waterhouse, Applied Science, London,1981.
9. Waterhous R B and Taylor D E, Fretting Fatigue in Steel Ropes, *J. Amer. Soc. of Lubric. Eng.* **27**,4, 123, 1971.

Chapter 10

The mechanisms and control of wear

D T Gawne – South Bank University, London

10.1 Background

Tribology concerns the performance of engineering components under sliding and rolling contact. The mechanisms of friction, lubrication and wear are described in terms of the micro-geometry of surface interactions. This chapter explains how an understanding of the theory can be used as a basis to solve tribological problems.

Tribology is the science and technology of solid surfaces in contact and in relative motion. It is the study of *friction, lubrication* and *wear* of engineering surfaces in order to understand the detailed surface interaction with the aim of achieving improvements in a specific application. The application is of central importance in tribology and the subject is thus very much an applied science. The underlying reason for this is that tribological performance is very sensitive to the operating conditions and environment: e.g. the type of contacting materials, the surface topography, the mechanical system, lubrication and atmosphere. Although the principles of tribology are general, the solutions are therefore particular to the application, whether this is extending the life of a twist drill or reducing skidding of automobile tyres on wet roads.

Tribology is widely recognised as a subject of major technological and economic importance. A series of reports [1–4] over the last 25 years has shown that approximately *1.5% of the gross national product* of an industrialized country could be saved by control of tribological behaviour. This represents a total saving for the UK of approximately *£5 billion per year* at the present time (1993). The savings derive from factors including maintenance costs, replacements, breakdowns, downtime, energy consumption and increased life of machinery and artefacts. For example, the annual output of lathes and milling machines alone in Germany is approximately £1.5 billion. A particularly important finding was that the first 20% of the above savings could normally be achieved without any significant investment. The US [2] and Chinese [3] reports estimated that the average investment/savings ratio was approximately *1:60* such that an expenditure of 1 unit on research and development would generate 60 units of savings.

Tribology offers many great challenges for the future [5–8]. For example, the tribological design of materials and lubricants for high temperature engines could lead to the *100 mile per gallon family car*. There are major possibilities in the field of medical engineering as exemplified by the fact that the original artificial *hip joints* had an expected life of only 3 years, whereas 25 years is now normal and further tribological research is anticipated to extend this to the life expectancy of the patient. Tribology can make a considerable contribution to *environmental control* by devising alternative uses or methods of disposal for waste industrial oils, solvents, cutting fluids and coolants in order to minimize water pollution. Tribology also has an important potential rôle on the *macro-terrestrial* level of plate tectonics: the relative movement of the jigsaw of rigid plates that make up the Earth's crust. The application of tribological principles in the future may lead to an improved understanding of earthquakes, volcanoes and other seismic activity, and possibly even to some degree of control.

10.2 Nature of surfaces

Friction and wear are properties usually detected macroscopically by the measurement of forces and mass changes. However, the mechanisms governing these properties take place on a microscopic level and a mechanistic understanding is crucial to the control of tribological behaviour. It is thus essential to appreciate the nature of surfaces at the microscopic level.

10.2.1 Topography

A surface of a solid is the geometrical boundary between the solid and its environment. Just as the surface of the earth appears perfectly smooth when viewed from outer space, all solid surfaces are rough on a microscopic scale even though they may appear flat to the naked eye. The surface contains hills and valleys which generally vary randomly in height and spacing, although machining operations tend to produce a periodic surface topography known as the *texture*.

The topography of a surface has a pronounced effect on properties such as wear, lubricity and adhesion, so that it is clearly necessary to identify means of quantifying it. The most readily measurable parameter for surface roughness is the R_a *value*; previously known in the UK as the *centre line average (cla)* or the *AA* value in the USA. The R_a value is the average height of the peaks above their centre line. Referring to Fig. 10.1, the R_a value is obtained by adding the shaded areas and then dividing the sum by the length l. It is also useful to define this parameter mathematically as the *mean value of the peak height y with respect to the horizontal distance x over the range* $0 < x < l$:

$$R_a = \frac{1}{l}\int_0^l |y|.dx \qquad [10.1]$$

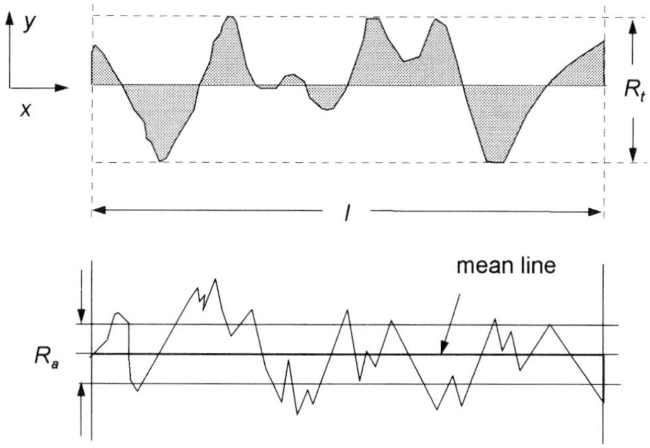

Figure 10.1 Schematic of a surface profile over a sample length of l.

Typical values of R_a vary from 0.05μm for polished surfaces, through 1μm for ground surfaces, 10μm for grit blasted and 30μm for heavily machined surfaces. However, the measurement of R_a does by no means completely describe the surface geometry as can be illustrated by Fig. 10.2 which shows a number of surfaces with the same R_a value but distinctly different profiles.

There are other height parameters in use. One is the R_t value which is the *height from the highest peak to the deepest valley found anywhere along a selected profile* (Fig. 10.1). The R_t parameter is sensitive to extremes and the position of the evaluation length but it is useful for applications such as seals, where a single scratch cannot be tolerated. A more representative parameter is the R_z value which is obtained by *averaging the five highest peaks and the five deepest valleys in the total traverse*; it is sometimes referred to as the "*ten point height*". Although the ratio of one parameter to another varies with the shape of the profile, to a

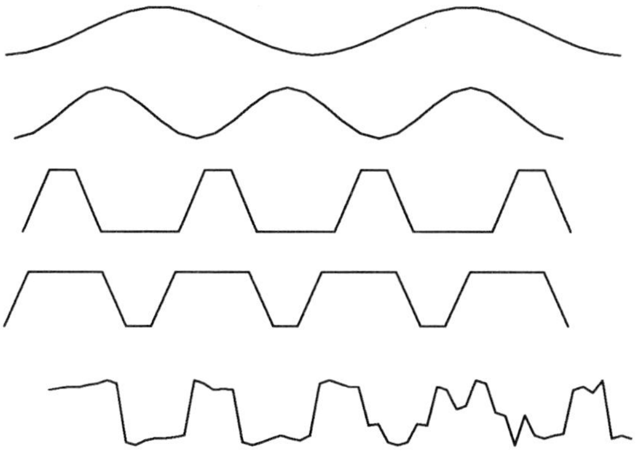

Figure 10.2 Surface profiles with the same R_a values.

rough approximation the R_z *value is five to ten times bigger than the* R_a. The R_t and R_z values both suffer from the same limitation as the R_a value in that surfaces can have the same R_t or R_z value but exhibit quite different profiles.

The horizontal component of the surface profile, the spacing of the significant peaks, may also have practical significance. There are various procedures for defining a *significant peak*: one is to count only those peaks for which the adjacent valley exceeds a given depth. Peak spacing clearly depends upon the definition applied but in terms of order of magnitude can vary from 0.5μm to 5mm. The height of the peaks is generally much smaller than their spacing and in order to portray a surface profile on a manageable small length of recording chart paper, it is usual to use a far greater vertical than horizontal magnification. The effect of this is to greatly exaggerate the sharpness or gradients of the sides of the peaks.

Surface finish and *texture* can be evaluated in a number of ways, the simplest being to run a fingernail over the surface. A light microscope provides useful information about the direction (the *lay*) and spacing of the surface features but little or none about their height. The scanning electron microscope gives sharp stereoscopic images of the surface but is generally limited to small areas and small specimens. Optical interference methods show contours of fine surfaces. The most widely used technique, however, is the stylus method which pulls a sharply pointed diamond tip (like the needle on a record player) over the surface in order to trace its profile.

Many parameters are required to define the geometry of a surface. In many practical cases, however, their measurement is too expensive and only the R_a value is obtained. It is important to interpret the R_a parameter with a full awareness of what it can and cannot tell about a surface. It is at best regarded as a practical index for comparing the heights of similar profiles on a linear basis: twice the $\mathrm{R_a}$ value, twice the height. In this regard, it has proved to be serviceable in situations where the same mechanical shaping operation (e.g. grinding, milling, blasting) is always applied, such as in process control.

10.2.2 Surface chemistry

The chemistry of the earth's surface is equally as complicated as its physical nature. The surface layers consist of deformed rock strata, covered by fine particulate solids such as soil or sand, with an outer layer of vegetation arising from the interaction of the surface material with the environment.

The examination of the surface of engineering material shows similar characteristics. For example, on top of the normal crystallized grain structure of a metal, there will be a layer of deformed and strain hardened material (to a depth of ~1–10 μm) created by the manufacturing process. This layer is often overlaid by a microcrystalline region (~0.1μm depth), often extremely hard, which is produced by the rapid cooling of the outer surfaces during manufacture. The outermost layer will be produced by chemical reaction of the surface with its environment and, in the case of a metal, will consist of a thin, often transparent, oxide film (0.01–0.1 μm thick) containing cracks and pores. Molecules of water oxygen and grease are weakly attached to the oxide. Finally, the surface will usually be covered with dust (particle size ~1 μm), wear debris and possibly lubricant. The characteristics of the outer surface may have considerable influence on tribological behaviour: the oxide layer may reduce wear and friction by acting as a barrier between two metal surfaces, while the dust particles may increase wear by scoring out the underlying material.

10.3 Contact mechanics

The previous section described how all surfaces are rough on a microscopic scale. Putting two solids together is rather like turning Switzerland upside down and standing it on Austria [5]. When two surfaces are placed in contact, they will only touch at the tips of a small number of peaks (*asperities*), whereas from a macroscopic view, the surfaces will be assumed to make contact all along the profile shown in Fig. 10.3. The *true* area of contact is thus much smaller than the apparent or nominal area.

Any load across the surfaces of the two solids is supported only by the places where the asperities make contact and thus only a very small fraction of the apparent area carries the load. This means that the local stresses at these asperities are much higher than expected from a macroscopic description of the apparent area. At very low loads, this causes the asperities to deform elastically. However, for realistic loads extensive *permanent or plastic deformation* takes place at the asperities tips until the true area of contact is increased to the stage at which it can support the load. Each contacting asperity therefore yields plastically, the true contact areas (as illustrated by a_1, a_2, a_3 in Fig. 10.3) increase until mechanical equilibrium is reached and the force of the applied load W is balanced by the strength of the supporting areas. This increased contact region is often called a *junction*.

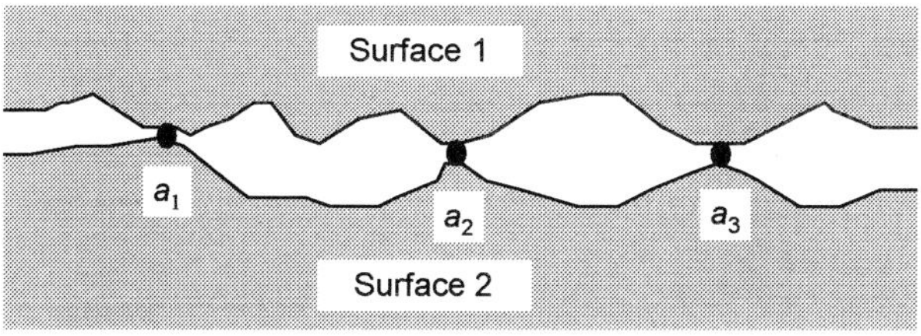

Figure 10.3 Schematic of *surface 1* pressed against *surface 2* under normal load W, showing that contact only occurs at asperities a_1, a_2 and a_3.

The mechanical equilibrium can be expressed:

$$W = a\sigma_y \tag{10.2}$$

where W is *applied load*, a is the *true area of contact* and σ_y is the *compressive yield stress of the softer surface*. The true area of contact is therefore given by:

$$a = \frac{W}{\sigma_y} \tag{10.3}$$

For example, if the load is doubled, then the true area of contact is also doubled. For many engineering materials, the yield stress is directly related to the *true hardness H* (*the load on a diamond divided by the projected area of the indentation*):

$$H = 3\sigma_y \tag{10.4}$$

The factor of the three derives primarily from the constraint on deformation by the material in the hinterland surrounding the indentation. *Equation 10.3* can therefore be re-written:

$$a = \frac{W}{3H} \tag{10.5}$$

where *H is the true hardness of the softer surface*.

10.4 Friction

A finite amount of horizontal force is required to slide one surface over another. Friction is the resistance to motion of one body sliding on another. *The force that will just cause sliding to start* (to initiate sliding), F_s, *is found experimentally to be directly proportional to the normal force or load W across the contact surface*:

$$F_s = \mu_s W \tag{10.6}$$

where the constant of proportionality, μ_s, is the *coefficient of static friction*. Once sliding starts, the limiting frictional force, F_k, decreases slightly and

$$F_k = \mu_k W \tag{10.7}$$

where μ_k ($< \mu_s$) is the *coefficient of kinetic or dynamic friction*.

The result that the friction between two sliding surfaces only depends on the force pressing them together and *not* on the area of contact seems surprising. For example, the frictional force between a brick and a plane is the same whether it stands on its small end face or its larger faces. In order to understand this behaviour, it is necessary to consider the micro-geometry of the surfaces.

The previous section showed that surfaces in contact only touch at the tips of certain asperities. At these regions of true contact, the atoms on one surface will attract those on the other to produce adhesion and form junctions. In the case of metals, this may be referred to as *cold welding* and is particularly marked in mechanical systems under vacuum, such as outer space. When sliding occurs, these adhesions have to be overcome, which in practice means that the junctions have to be sheared. The force to shear the junctions is the primary cause of friction between clean surfaces. The *frictional force* F *to shear the junctions* is thus:

$$F = a\tau_y \qquad [10.8]$$

where a is the *true area of contact* and τ_y is the *shear yield stress* of the junction, which is comparable with the bulk shear yield stress of the softer of the two materials in sliding contact. The shear yield stress is directly proportional to the compression yield stress, σ_y, and it can be shown that:

$$\tau_y = \frac{1}{2}\sigma_y \qquad [10.9]$$

combining equations [10.3, 10.8 and 10.9] gives:

$$F = \frac{1}{2}W \qquad [10.10]$$

which is a statement of the law of friction given in equation [10.6] for which $\mu = 0.5$. This is the correct order of magnitude for the friction between dry materials. A consideration of the micro-geometry of surfaces thus enables an understanding of the law of friction.

There is an additional cause of friction: if one surface is harder than the other, the roughness or asperities on it will *plough out grooves* on the softer which clearly requires an additional force. For clean surfaces, however, the *adhesion* component generally dominates, although under lubrication the *ploughing* component may be an appreciable fraction of the total frictional force.

In practice, the coefficient of friction is not constant at a value of 0.5 but can vary appreciably. There are various reasons for this including oxide layers, contaminant films, strain hardening and junction growth.

10.5 Lubrication

In many instances, high friction is undesirable: it is a process that essentially absorbs energy and so reduces the efficiency of machines. In addition, the work of friction is mainly converted to heat and the resulting temperature increases can damage or even melt the sliding surfaces. There are some cases, however, where *high friction is necessary such as in brake pads, clutch linings and car tyres.*

In order to minimize friction, it is necessary to make it as easy as possible for surfaces to slide over one another. The mechanism of friction indicates that this can be achieved by contaminating the asperity tips with a *barrier film* that: firstly, can *prevent contact* between the two surfaces thereby suppressing the formation of adhesive junctions; secondly, the film should have a *low shear strength* so that any junctions that do form can be easily broken. The film also clearly needs to be able to withstand the high pressures at the sliding surfaces. The ideal way to achieve these conditions and low friction is to run the surfaces under *hydrodynamic* or *fluid film lubrication*, where the thickness of the oil film is much greater than the height of the asperities and so no contact takes place between the two solid surfaces. In this case, the chemical properties of the oil are of little importance, only its viscosity characteristics are crucial. Typical values of the friction coefficient under hydrodynamic lubrication are 0.001-0.005.

An example of hydrodynamic lubrication occurs in *journal bearings* in which a round shaft or journal, immersed in oil, rotates in a cylindrical bearing (Fig. 10.4). This arrangement is widely used in many types of rotating or reciprocating machinery, such as the crankshaft bearings of an automobile. The oil is viscous and so the revolving shaft drags the oil around with it (Fig. 10.5). As a result, the oil is squeezed into the converging gap between the shaft and the bearing and enormous pressures generated (*up to* 10^7 *Pa*) are enough to force the shaft and the

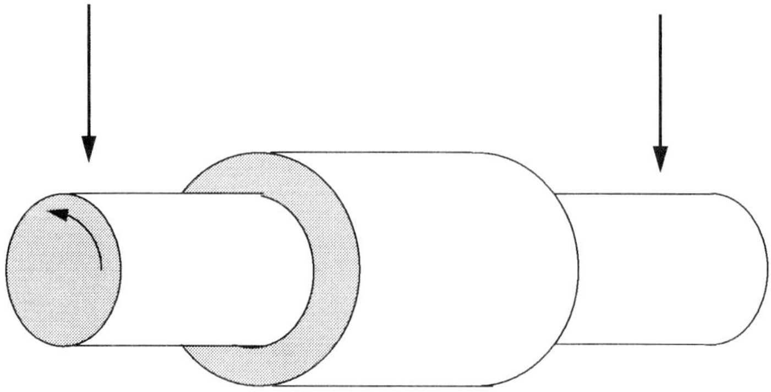

Figure 10.4 Journal bearing.

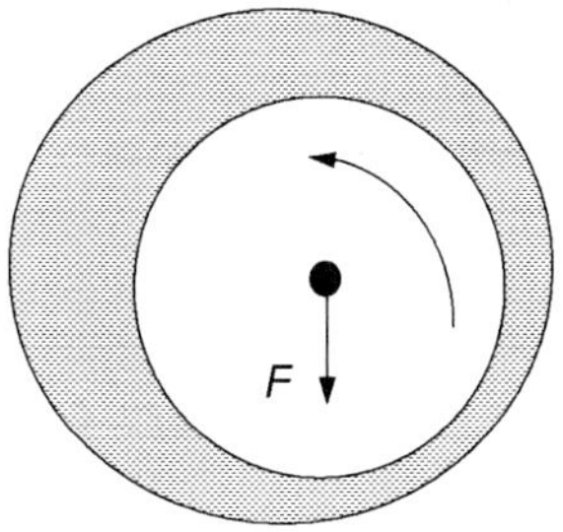

Figure 10.5 Cross-section through journal bearing showing the convergent oil film between the revolving shaft and bearing.

bearing apart: there should be no asperity contact and no wear under ideal conditions. However, the high pressure and load capacity can only be generated at high rotational speeds of the shaft and fairly high viscosity oils. When starting up an engine or running slowly under high load, hydrodynamic lubrication is not present and the less effective *boundary lubrication* (as explained below) has to be relied upon. Under these conditions some contact and wear of the mating surfaces will occur. This is one of the reasons why car engines do not last as long when used for short runs compared with long journeys.

If the oil film thickness is less than the height of the asperities, then contact between the two sliding surfaces can occur through the liquid film. It is very beneficial under these conditions to add small amounts (~1%) of active organic chemicals to the oil. For example, *linear chain molecules with one polar end* are

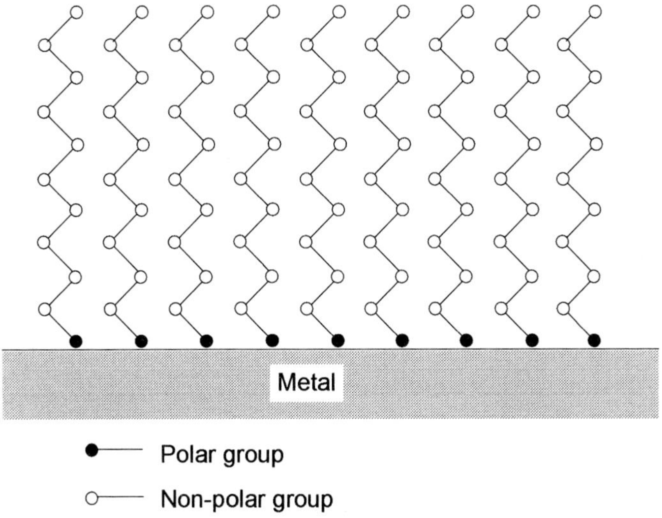

Figure 10.6 Schematic of adsorption of polar molecules on a solid surface.

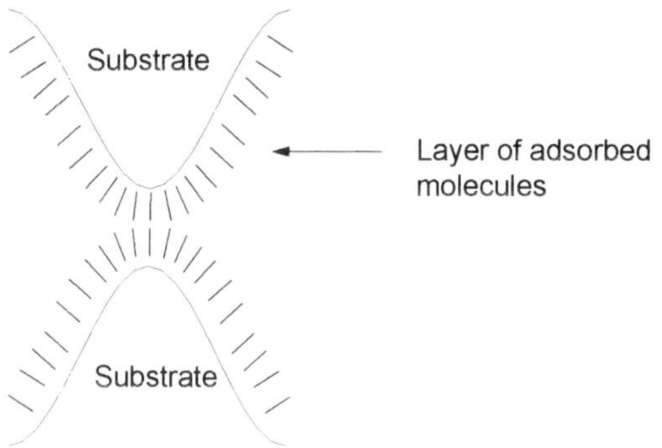

Figure 10.7 Schematic showing how adsorbed active molecules can act as a barrier layer between solid surfaces.

highly desirable because the polar end is attracted to a metal or metal oxide surface and sticks to it. The rest of the molecule is attracted to its neighbouring molecules to form an oriented "*carpet*" of molecules as schematically shown in Fig. 10.6. These carpets can resist very large forces normal to the surface and hence separate the asperity tips very effectively (Fig. 10.7), whereas the two layers of molecules can shear over themselves easily. This mode of lubrication is called *boundary lubrication* and is encountered in applications such as door hinges and machine slideways. It is capable of reducing the friction coefficient by a factor of 10 with typical values for metals in the range 0.05–0.1.

10.6 Wear

Wear is the damage caused by contacting surfaces in relative motion, usually sliding or rolling under load. In many cases, wear can be quantified by the loss of material from a surface, although some materials can wear without the removal of surface material. For example, scratching the surface of ductile copper clearly produces wear but it often occurs by plastic deformation producing micro-displacement of material to form grooves and no fracture nor loss of material need take place. Nevertheless, wear is commonly defined for convenience as *the loss of material from contacting solid surfaces in relative motion*. The loss of material is usually progressive rather than catastrophic and the amounts are often quite small. For example, although breakage and obsolescence in an automobile are clearly visible, wear may be undetectable by casual inspection. In fact, a modern 1000kg automobile can be completely worn out when much less than 1kg has been worn off the sliding surfaces. In general, the most important mechanisms are *adhesive, abrasive, corrosive and fatigue wear*.

10.6.1 Adhesive wear

In theory, two clean surfaces of similar crystal structure should adhere strongly to each other when they are simply placed in contact due to the forces of attraction between the atoms on each surface. In practice, this does not usually happen because of surface contamination and because the atoms on one surface are not brought close enough to those on the other surface for the short-range attractive forces to act. However, if the surfaces are pressed together by a normal load and moved by sliding one surface over the other, then the surface film breaks up, the atoms are brought into close proximity and appreciable adhesion can occur.

Figure 10.8 shows schematically the adhesion between two asperities on the surfaces of materials A and B. If fracture takes place at the initial sliding interface, no wear will occur. However, if the adhesion between A and B is sufficiently high, then fracture will occur away from the sliding interface and transfer of one material to the other will take place resulting in adhesive wear. In general, the fracture will occur in the softer material A, which will then be adhesively transferred to the surface of the harder material B.

The mechanism of adhesive wear outlined above crucially depends upon the magnitude of the *true area of contact*: adhesion is only possible where the surfaces are touching. If it is assumed that the wear particles or debris (e.g. resulting from the transferred fragment of A in Fig. 10.8) are all geometrically similar, then the volume of wear would be expected to be proportional to the true area of contact, at which adhesion occurs, and also the distance of sliding, L. The true contact area, a, has been given previously by equation [10.5] as:

$$a = \frac{W}{3H} \qquad [10.5]$$

where W and H are the *load and hardness of the softer material* respectively. The volume of wear, V, will be proportional to product of a and L so that

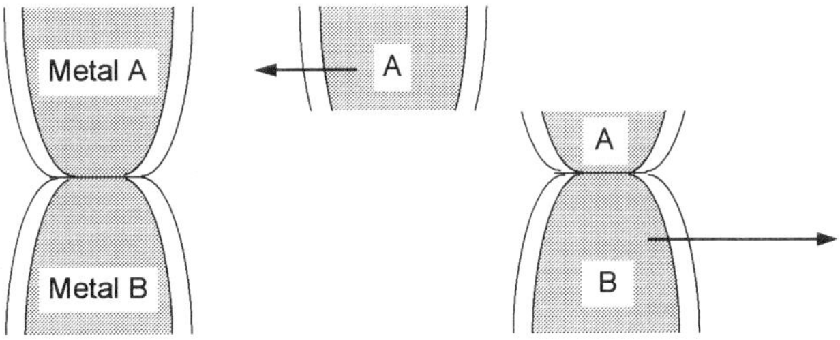

Figure 10.8 Adhesion between two asperities giving rise to transfer of metal A to the surface of metal B.

$$V = kaL \qquad [10.11]$$

where k is the constant of proportionality and L is the *sliding distance*. Combining equations [10.5] and [10.11]:

$$V = k\frac{WL}{H} \qquad [10.12]$$

Equation [10.12] is sometimes referred to as the *Archard equation* and k as the *adhesive wear coefficient*. The equation clearly applies to idealized conditions but provides a very useful preliminary model for appreciating adhesive wear. Equation [10.12] implies that wear is proportional to the sliding distance, the applied load and the hardness of the softer material. The wear coefficient k may be interpreted as the *probability of wear particles being created during sliding of the two surfaces*. Generally, if two materials are mutually soluble, such as iron and nickel, their atoms will tend to attract strongly and adhere giving rise to high wear coefficients and wear rates. Conversely, a metal like lead has a good adhesive wear resistance against iron and a wear coefficient two orders of magnitude smaller than nickel-iron because the metals are mutually insoluble.

10.6.2 Abrasive wear

Abrasive wear arises from the cutting action of hard surfaces sliding on softer materials, as for example, when hard surface asperities act like cutting tools and remove material from the softer surface. It is similar to making a hardness indentation with a diamond indenter in a material and then moving the indenter sideways to plough out a groove. Figure 10.9 shows an idealized schematic of a single conical asperity ploughing out a groove in a softer material. Another important mechanism leading to abrasive wear arises when a loose *debris particle* is trapped between the sliding surfaces: the loaded particle penetrates the softer

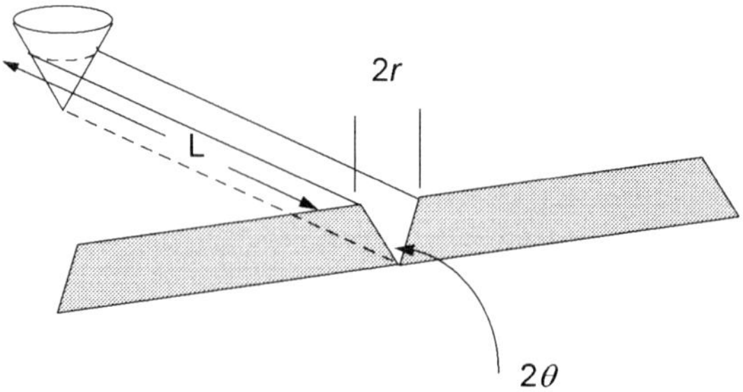

Figure 10.9 Abrasive wear due to a conical asperity ploughing out a groove in a softer material.

surface and scratches out a wear groove. Such debris may be extraneous, such as sand particles, or be formed in-situ by the primary wear as oxide particles.

A useful equation representing abrasive wear can be obtained by considering the action of the single conical asperity in Fig. 10.9. This asperity slides a distance of L and displaces volume of material $Lr^2 \cot\theta$. If the compressive yield stress is σ_y, the normal load carried by this asperity is $\pi r^2 \sigma_y/2$, the factor ½ arising because only the front part of the asperity makes contact with the soft material. If there are n such asperities between the surfaces, the total load W is given by:

$$W = \frac{1}{2} n \pi r^2 \sigma_y \qquad [10.13]$$

and the total volume of material removed by the n asperities is:

$$V = nLr^2 \cot\theta \qquad [10.14]$$

Eliminating nr^2 from these equations gives:

$$V = 2\cot\theta \frac{WL}{\pi\sigma_y} \qquad [10.15]$$

From equation [10.4], $\sigma_y = H/3$, where H is the hardness, and so the wear volume V is given by:

$$V = k\frac{WL}{H} \qquad [10.16]$$

where k is the *abrasive wear coefficient* and is given by:

$$k = \frac{6}{\pi}\cot\theta \qquad [10.17]$$

Equation [10.16] is of the same form as that for adhesive wear in equation [10.12]. The difference being that the wear coefficient is now related to the geometrical characteristics of the hard asperity as defined by the angle describing sharpness. In both cases, the wear coefficient may be determined from the gradient of a plot of wear rate (V/L) against load. It must be pointed out, however, that in some practical cases the wear rate is *not* a linear function of load and so a general wear coefficient *cannot* be defined.

The wear rate of a material has been found [9], not surprisingly, to decrease substantially as its hardness, H, approaches that of the abrasive, H_a, particularly for $H/H_a > 0.8$. Silica is the most common abrasive in the form of dust, grit, sand, rock or gangue and has a Vickers hardness of ~6.5 GPa, ~10 MPa is equal to one *Vickers hardness number*. In many cases, therefore, it would appear to be unnecessary to use hardnesses above this value, especially since a reduced fracture toughness generally accompanies high hardness. Nevertheless, there are harder abrasives than silica and each application needs to be considered on its own merits.

The wear equations given in equations [10.12] and [10.16] often provide useful insight into wear phenomena. However, they are *simplifications* of reality and other properties, such as fracture toughness and Young's modulus, also make a significant contribution to tribological behaviour.

10.6.3 Corrosive wear

Corrosion is the degradation of a surface by chemical reaction with the environment. A clean surface of a solid, such as a metal, will generally react with its environment and the resulting corrosion product will form a contaminant film. The rate of formation of such films is initially very rapid but then gradually decreases as the film thickness and diffusion distances increase.

In many cases, such as rust on steel, the surface film adheres relatively loosely to the surface. A rubbing action may thus scrape off the film exposing the underlying clean solid, which immediately reacts with the environment to produce new surface films that are again removed during further rubbing. Material is therefore being continuously removed from the surface and a form of wear is taking place: *corrosive wear*. The chemistry of these mechanically stimulated reactions is influenced by factors such as local frictional heating, residual stress, humidity, load and speed. *Oxidative wear* is a common subset of the more general phenomenon of corrosive wear, since oxidation may be considered as dry corrosion in the absence of water.

The fragments of film being removed by the rubbing process become part of the *wear debris*. In the case of oxide films, the debris particles may cause *three-body abrasive wear* by becoming trapped between the sliding surfaces, since metal oxides are generally much harder than the metal itself. For example, *aluminium* is a relatively soft metal (Vickers hardness ~200 MPa) whereas aluminium oxide is very hard (Vickers hardness ~18 GPa) and, in fact, is often used as the cutting agent in grinding wheels.

A further effect of a corrosive environment may be to enhance the abrasive action of the wear debris. Metallic debris particles can react with the environment to produce extremely hard oxides and a much increased rate of abrasive wear on the sliding surfaces.

Corrosive environments may have beneficial effects on wear. For instance, a mechanically stable oxide film that does not flake off during sliding can provide an effective *barrier layer*, prevent metal-to-metal contact and reduce the wear rate. In order to form a mechanically stable film, the oxide must adhere strongly to the metallic substrate, have a limited thickness for flexibility and similar mechanical properties to the metal. The similarity in properties is necessary to enable the oxide film and the underlying metal to deform together under load without break-up of the film. For example, tin sliding against itself gives rise to high friction and wear (Vickers hardnesses of tin metal and tin oxide are ~100 MPa and ~16 GPa respectively), whereas chromium sliding against itself produces low friction and wear (Vickers hardnesses of chromium metal and chromium oxide ~10 GPa and ~20 GPa respectively).

In certain applications, surface films are artificially produced in order to reduce friction and wear. *EP (extreme pressure)* additives to lubricating oils produce surface films of e.g. chlorides or sulphides, which act as protective layers on metallic components: the applied load is supported through the soft surface film by the base material, while any shear takes place within the film. The films are particularly valuable in high pressure contacts such as hypoid gears as used in automobile back axles.

10.6.4 Fatigue wear

Fatigue wear may result when a surface is loaded cyclically due to repeated sliding, rolling or impacts. Fatigue failure can occur after a larger number of loading cycles, even though the load is less than that required to cause failure in a single load application. However, the lower the applied cyclical stress level, the longer the life of the material before failure occurs. Wear processes, in general, involve the repeated stressing and unstressing of a material, which initiates cracks at or just below the surface. Further cycles cause the cracks to grow, link up and eventually break through to the surface to produce discrete wear debris particles. As a result of this time-dependent mechanism, the surface appears unaffected for long periods before pitting or spalling suddenly occurs followed by rapid wear. Experimentally, it is generally found that the *lifetime or time to failure*, t, is markedly dependent on the applied load, W, and for the roller cylinder bearings:

$$t = \frac{C}{W^2} \qquad [10.18]$$

where C is a constant.

All that is required to produce the above fatigue failure is that the surface material should be cyclically loaded, whereas the adhesive and abrasive wear mechanisms require direct physical contact between the solids as well as loading. If the surfaces are separated by a lubricant film and abrasive particles excluded, then the adhesive and abrasive wear mechanisms *cannot* operate. However, the applied load is *still* transmitted to the solid surface through the lubricant film causing surface stresses and the possibility of significant fatigue wear. This is particularly true in rolling contacts, where the stresses are high and the slip is small. Ball and roller bearings, gears and cams are examples where a fatigue wear mechanism is commonly observed and gives rise to pitting or spalling at the surface. The debris particles tend to be much larger than those generated by adhesive/abrasive wear (e.g. 1mm compared with 30μm) and may become trapped between the moving surfaces to cause significant abrasion.

10.7 Wear transitions

Equations [10.12] and [10.16] derived from the simple models of wear indicate that the wear rate should linearly increase with the applied load. In practice, this is often *not* the case and transitions may occur in which the wear rate suddenly increases rapidly from *mild* to *severe* as a result of only a small increase in load or speed (Fig. 10.10).

In *mild wear*, the worn surface is smooth, the debris particles are small and often oxidized, and the wear rate is low. In *severe wear*, the surface is rough, deeply torn, often uniformly distributed, the debris particles are large and often metallic, and the wear rate is high. When the contact pressure and sliding velocity are both low for steel, a thin, tough oxide forms, which prevents metal-to-metal contact and provides a low wear rate. The mechanism of material removal is corrosive or oxidative wear. Just above the first transition load (Fig. 10.10), the oxide layers are penetrated and metal-to-metal contact occurs giving *severe adhesive wear*. At yet higher loads above the second transition (Fig. 10.10), a hard *martensitic* surface layer is formed on the steel surface giving a low wear rate. The layer forms because of the high flash temperatures generated at asperities due to the frictional heating followed by rapid quenching as the heat is conducted

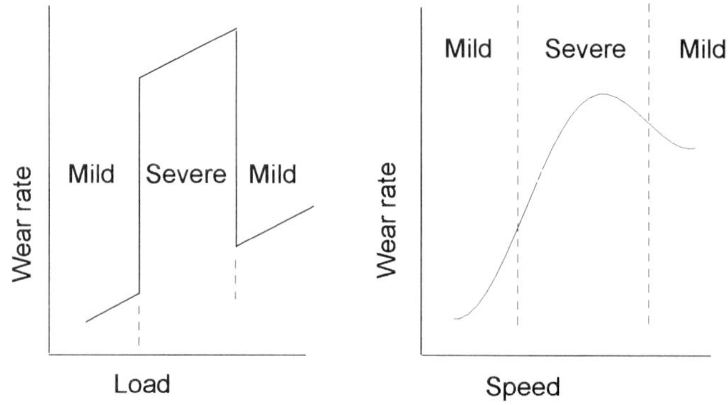

Figure 10.10 Schematic representation of load-dependent and speed-dependent transitions in wear rate.

into underlying metal. The first transition at high sliding speeds has been attributed to frictional heating producing local temperatures above the melting point of steel causing a liquid film to form at the contact and resulting in *severe wear*. At yet higher speeds, the increased interfacial temperature causes the thick oxide film to deform plastically (or even melt-flow) while also thermally insulating the underlying metal, which results in a low wear rate and the second transition with the sliding speed (Fig. 10.10). Distance-dependent transitions in wear rate are also commonly found in sliding systems. It is observed in the initial stages of high wear or *running-in* of sliding surfaces. The running-in period is generally attributed to the presence of high spots on the surface profile and misalignments between the two sliding surfaces, which lead to high local pressures and high interfacial asperity temperatures. As sliding progresses, the high peaks and misalignments are removed, the *true area of contact* increases and the true contact pressure decreases. In addition, the surface layers strain harden due to the plastic deformation associated with sliding contact, patches of martensite may form and protective oxide films develop leading to a substantial reduction in wear rate. Two régimes thus often exist with respect to sliding distance: an initial running-in (or *transient*) period with a high wear rate followed by a transition to an equilibrium wear stage with a much lower wear rate. This explains why it is often expedient to apply low loads and speeds to new machinery and engines at the beginning of their service life to allow them to *run in* without serious damage.

Wear transitions can be extremely important in practice since a small variation in operating conditions of a machine may suddenly produce a dramatic increase in wear. It is clearly necessary to test the materials, lubricants and mechanical systems *over the entire* possible operating range of load, speed and sliding distance to determine whether or not there is a risk of a wear transition.

10.8 Control of wear

The traditional methods of controlling wear include the use of rolling elements, lubrication and design of the mechanical system. However, the previous sections indicate that there may also be solutions based on *materials selection*. Since wear is essentially a surface phenomenon, a localized treatment of the surface is often a much more attractive solution economically than manufacturing the whole component from a wear resistant material. *Surface engineering* may also present the best technological solution as the ideal wear-resistant material may not be suitable for manufacturing the whole component. For example, a ceramic may be ideal for wear resistance but is often excessively brittle in the form of a monolithic component; a thin coating of a ceramic on a metallic substrate may offer high wear resistance together with acceptable toughness.

The previous section shows that high abrasive wear resistance is obtained by using materials with a high hardness (equation [10.16]). *Electroplated chromium* grades have hardness values in the range ~7.5 to 9.5 GPa and thus have major potential as protective coatings against abrasion. Chromium plating is used widely as a protective coating on steel: 1–10μm coatings for drills, 10–200μm coatings for drawing dies, 10–75μm coatings for pump shafts and 10–100μm for hydraulic rams. The deposit thickness varies with the application based on practical experience, although there is also evidence showing that the coating life increases with increasing coating thickness [6].

The hardness of *electroless nickel* coatings can be doubled by heat treatment from ~6 GPa in the as-deposited state to ~11 GPa after heat treatment for 1 hour at 400°C. It is frequently assumed that heat treated electroless nickel has at least the same wear resistance as chromium plating because of its higher hardness. This is *incorrect* and results show that chromium has the superior tribological properties [7]. The reason for this lies in the wear mechanism: the dominant wear mechanism when electroless nickel slides against many metals is *adhesive wear*. This is primarily due to the mutual solubility of nickel and the other metal such as iron; adhesive wear can be considerably reduced by heat treatment since this transforms the amorphous nickel mostly into nickel phosphide, which is insoluble in iron. In cases of heavy gouging or applications requiring flexibility, however, heat treatment may adversely affect wear resistance as it reduces the fracture toughness. Nevertheless, electroless nickel provides moderate wear resistance and is satisfactory for many applications.

The high wear resistance and low friction coefficients of *chromium plating* are attributed to the thin passivating, self-healing oxide (Cr_2O_3) layer on its surface as well as its high hardness. The Cr_2O_3 film acts as a barrier layer preventing contact between the two sliding metals, suppressing adhesive transfer by presenting an incompatible, insoluble surface to a steel counterface. The high hardness of the chromium is also beneficial in that it provides, for example, a firm supporting base for the Cr_2O_3 film and thereby minimizes the development of tensile stresses at the Cr_2O_3/Cr interface during wear and the consequent risk of film failure.

There is a wide range of techniques available for depositing hard coatings for wear resistance of components. *Flame spraying, high velocity oxy-fuel spraying* and *plasma spraying* can be used to produce a wide range of ceramic and ceramic alloy coatings. For example, the addition of titania to the alumina is found to improve the fracture toughness and wear resistance of the coatings. Further improvements in fracture toughness and wear resistance for some applications can be obtained by using cermet composite coatings, in which the metal matrix acts as a crack stopper. An example is cobalt–tungsten carbide produced by high velocity oxy-fuel spraying.

Direct evaporation, sputtering, chemical vapour deposition are further processes by which hard coatings (e.g. carbides, nitrides or oxides) can be deposited. *Physical vapour deposition of titanium nitride* coatings is now widely used to enhance wear resistance. The process involves evaporating titanium in low pressure nitrogen/argon atmosphere to form titanium and nitrogen ions, which are attracted to the negatively biased workpiece where they deposit as titanium nitride. A major application is the coating of cutting tools used for the manufacture of gears, *where tooling life increases of five to ten times* are common with important savings in material costs and downtime. Similar increases in life are obtained by coating tools and punches with titanium nitride for use in sheet metal forming for *only 15%* increase in the total price of the tool. An application that may become more important in the future concerns titanium alloys, which have scope in engines and machine parts for transportation owing to their strength and lightness. However, their application has been seriously limited because of titanium's poor resistance to abrasive wear. Titanium nitride coating and *ion implantation* have both been shown [8] to provide considerable improvements in wear resistance, Titanium nitride coatings are now well established in Formula one racing cars where they are used in parts for gearboxes, wheel hubs, differentials and steering racks.

An alternative approach to the problem of wear is to produce *soft coatings rather than the hard deposits* described above. This involves coatings of such materials as *lead, indium and PTFE*. These materials interpose between the asperities of the sliding materials and reduce the opportunities for adhesion. They also have low shear yield strengths so that the stress to break junctions at the asperities is small. The integrity of the coating is important: it must adhere strongly to the substrate and be of sufficient thickness under conditions of abrasive wear to embed hard debris particles (e.g. in-situ wear debris or extraneous grit) so that they are not available to score the counterface. Aluminium-tin bearings may be coated with *indium or lead* to reduce friction and to allow some accommodation of misalignment of the shaft. *PTFE* has a low surface energy with one of the lowest friction coefficients of all polymers and is widely used as an anti-wear coating.

Laboratory wear tests provide *relative rather than absolute data*. The value of laboratory work is as compatibility tests in which the number of possible material combinations can be reduced to manageable proportions. An essential initial stage in the *materials selection* process is *wear diagnosis*, which involves a detailed examination of the worn components in the particular application in order to identify the predominant wear mechanism. A range of materials is then tested on laboratory apparatus that reveals the relevant wear mechanism. On the basis of the relative wear rates obtained, a provisional material combination is chosen, prototypes are tested on a pilot machine, and then in-service trials are carried out before the final materials selection is made.

Materials selection for wear applications is particularly sensitive to the details of the specific case under consideration. This may be illustrated by considering the tribological behaviour of a journal rotating in a bore lined with a white metal bearing surface. Detailed examination showed that the journal wears against the much softer white metal by the abrasive action of hard debris particles (e.g. oxides) embedded in the soft metal bearing surface. This necessitates the selection of a journal material with a high abrasive wear resistance. However, in some applications, the more critical requirement for the journal material is a high wear resistance against the bearing shell material (the substrate of the white metal layer), since under emergency conditions (e.g. the failure of the oil pumps and loss of lubrication), the white metal layer may be removed. This places *different* requirements on the journal material and high wear resistance against the bearing shell metal is now required, which *modifies* the basis of the materials selection.

References

1. "*Lubrication (Tribology) Education and Research*", Jost Report, Department of Education and Science, HMSO, London, 1966.
2. "*Strategy for Energy Conservation Through Tribology*", American Society of Mechanical Engineers, New York, 1977 and 1981.
3. "*An Investigation on the Application of Tribology in China*", Report by the Tribology Institution of the Chinese Mechanical Engineering Society, September 1986, Beijing.
4. "*A Strategy for Tribology in Canada*", NRCC 26536, National Research Council, Canada, 1986.
5. Bowden F P and Tabor D, "*The Friction and Lubrication of Solids*", Part II, Oxford University Press, 1964.
6. Gawne D T, *Thin Solid Films*, **118**, p385, 1984.
7. Gawne D T and Ma U, *Surface Engineering*, **4**, 239, 1988.
8. Garside B and Sanderson R, *Metals and Materials*, **7**, 165, 1991.
9. Richardson R C D, *Wear*, **11**, 245, 1968.

Chapter 11

Methodology of wear testing

C Subramanian – The University of South Australia

11.1 Introduction

Tribology, which encompasses friction, lubrication and wear, plays an important rôle in the economy of the industrialised world. A better understanding of its principles can improve the quality of products, efficiency and productivity. *Wear testing methodology* is important in the selection of materials for particular applications. *Surface engineering* involves various surface treatments and coatings to improve performance related to wear, corrosion, fatigue, friction and bio-compatibility. Many surface engineering techniques used in industry are enhancing the performance of tools and components in wear and/or corrosion situations. In this Chapter, various surface treatments and coatings, wear types and testing methods available for evaluation of tribological response of materials, especially the surface engineered are discussed. The selection of wear testing methods, measurement of wear rate characterisation of worn surfaces and debris for possible determination of wear mechanism(s) and any possible correlation of laboratory tests to those of field trials are highlighted. Various case studies involving simulation of particular applications are also described.

A standard definition for tribology is "*the science and technology concerned with interacting surfaces in relative motion*" [1]. It originates from the Greek word *tribo* which means "*rubbing*". It embraces the study of friction, wear and lubrication.

Tribology is truly an interdisciplinary area of research which is of interest to physicists, chemists, material scientists and engineers. *Friction* has been a main concern of physicists, whereas *lubrication* has largely been a research domain of chemists. The third aspect of tribology *wear* has been dealt with by mechanical engineers. However, in the last thirty years or so the systematic study of wear attracted material scientists and metallurgists. This might perhaps be due to the inability of the mechanical engineers to "see" beyond the continuum concepts of materials and to the ability of the metallurgists to recognise the importance of microstructure and structure/property relationships in general. The material scientists' contribution to wear has been significant at it is evident from the published literature concerning the rôle of microstructure on the tribological properties of materials.

Wear is a complex phenomenon as many factors influence it. A systems approach has been suggested to study wear and its related constituent elements [2]. A typical wear system is shown in Fig. 11.1 with input and output parameters. Considering the number of variables in the system, it is not surprising that wear data collected under well-controlled laboratory conditions cannot be directly correlated to those obtained in the actual field. Another reason for poor correlation between laboratory and field test data is the concurrent occurrence of two or more wear mechanisms in the system under actual conditions of application.

There is no formal classification of wear which has been fully accepted by researchers in the field of wear. However, to deal with a complicated problem such as wear, we need to adopt some classification based on one or more criteria. One of the most widely used wear classifications is due to Burwell and Strange [3] which is mainly based on four types of wear: *abrasion*, *adhesion*, *corrosion* and *fatigue*. Though these four types appear to uniquely represent various wear phenomena, there are a few drawbacks:

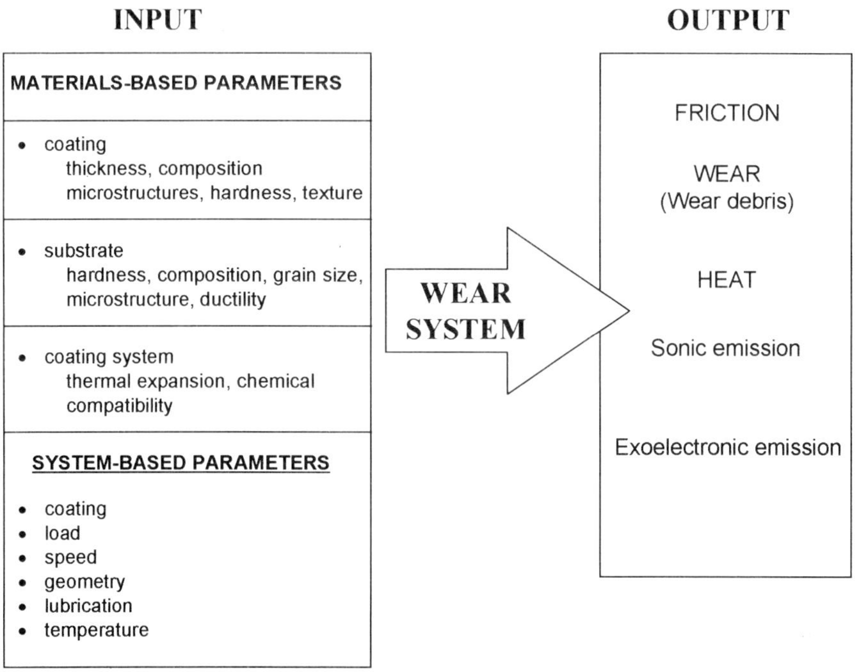

Figure 11.1 A typical wear system showing various input and output parameters.

Table 11.1 Classification of wear

Basis of classification	Wear types (e.g.)
1. General	Adhesive, abrasive, corrosive, fatigue
2. Motion	Sliding, reciprocating, impacting
3. Geometry	Pin-on-disc, pin-on-ring, cylinder-on-cylinder
4. Mechanism	Delamination, oxidation, seizure, adhesion
5. Applied load	Low stress, high stress
6. Lubrication	Lubricated (boundary, mixed, hydrodynamic), unlubricated (dry)
7. Material	Metal, ceramic, polymer, composite

i. It is not complete,
ii. it does not describe the production of wear particles from the surface and eventual ejection out of the wear system, and
iii. two or more mechanisms of debris production operate simultaneously or successively.

A recently developed classification of wear is due to Lancaster [4] who advocates material-specific wear types: metals, ceramics and polymers. These and other classifications are listed in Table 11.1. It is therefore advisable to mention the actual condition by stating the type of geometry, lubrication, materials involved and *predominant mode(s) of debris production*, rather than just stating an assumed wear mechanism (e.g. adhesion, delamination, etc.). Any wear study should concentrate on the nature of the worn surfaces and the wear debris in order to shed some light on the wear mechanism(s) of debris particle formation and eventual release from the wear system. If the wear particles formed in a wear system are not removed, they will lead to different wear mechanisms such as a three-body situation which has been extensively studied by Codet [5]. This is the very reason we use oil filters in our cars!

11.2 Surface engineering

Surface engineering has been defined as "*the application of traditional and innovative surface technologies to engineering components and materials in order to produce a composite material with properties unattainable in either the base or surface material*" [6].

One of the main aims of surface engineering is to improve properties such as friction, wear, corrosion, fatigue and bio-compatibility. The basic idea of surface engineering is to use leaner (often cheaper) materials as substrates and rare or strategically important (often expensive) materials as coatings.

The traditional methods such as carburising, flame hardening and shot peening and the newer ones such as physical vapour deposition (PVD), chemical vapour deposition (CVD) and laser processing, have been successfully employed to improve the performance of many engineering components in service. The thicknesses of various coatings or the depths of surface-modified material are compared in Fig. 11.2 [6]. It is obvious from Fig. 11.2 that the layer could be as thin as a few atomic layers in ion-implantation to as thick as a few mm in flame sprayed/weld-deposited coatings. The processing temperature can vary from sub-zero (e.g. cryogenic treatment) to a temperature close to the melting point of the substrate (e.g. laser cladding).

11.3 Wear testing

Wear tests are carried out for various reasons:

a. to evaluate resistance of a material or a pair of materials to wear,
b. to evaluate the performance of lubricants,
c. to find out the friction coefficient between two materials,
d. to understand the fundamentals of tribology (generally under well-controlled conditions between pure metals, and
e. to estimate the lifetimes of components and tools used in engineering and other applications.

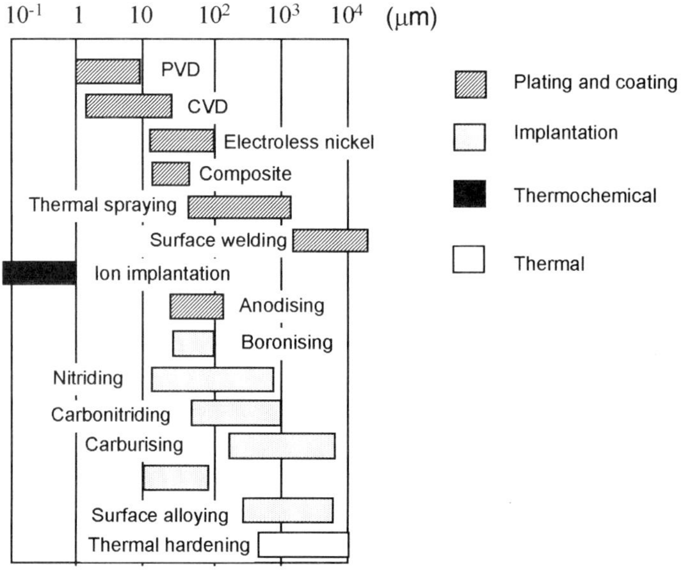

Figure 11.2 Thickness of coatings or depths of surface treatments.

The methodology of wear testing has been found in published literature, notably in the *Special Technical Publications* by *ASTM* [7–10]. Among the ASTM books on wear testing, there is one for coatings [9]. This book deals with many coatings such as plastic coatings and electroless nickel coatings. However, it fails to discuss the special problems encountered with wear of hard thin films.

Wear testing can be done in many ways: from a simple test in a laboratory to a test in the actual field. The cost involved in a laboratory test is very small when compared to a test in the actual environment. However, the laboratory test data may not be useful unless compared and correlated to those obtained in the field tests. Often, it is found that these sets of data are far apart, unless due care is taken in the laboratory to simulate the actual field conditions. Generally, the load and speed levels used in the actual conditions are maintained in the laboratory test. Other variables, such as the geometry, the counterface and the size of the specimens may only be occasionally duplicated in the laboratory. When everything is duplicated, the laboratory machine will be a copy of the actual rig! It is worth considering various stages of wear testing [11]. The level of simulation of the wear test can be selected depending on the element of risk and closeness of the actual application desired.

There have been many studies done on wear behaviour of coatings and surface-modified materials in laboratory testing machines and actual field conditions. However, there are only a few investigations which relate these sets of wear data observed in laboratory to the field data.

11.4 Coatings

Coatings or surface treatments should be fully characterized before any tribological testing is performed. There are many coating evolution tests available [12]. Among them, the techniques to measure thickness, adhesion and hardness of the coatings or surface-modified materials are considered to be important. Recently Buckle [13] has considered the requirements of an ideal quality assurance test. It should be:

- non-destructive,
- easily adaptable to routine testing,
- simple to perform and interpret,
- quantitative,
- applicable, and
- amenable to automation.

As there is no single test which meets the above mentioned criteria, it has been suggested that *scratch testing* and *hardness testing* would form bases for quality assurance tests. In the following two sections we will briefly discuss hardness and adhesion of coatings.

11.4.1 Hardness

Depending on the thickness of the coating or of the surface-modified layer, the method, load and the orientation of hardness measurement are determined. To obtain a "*true*" value for the hardness of the coatings, it is necessary to have a thicker coating (generally ten times greater than the depth of indentation) so that the indentation is not influenced by the coating/substrate interface or free surfaces [13]. This is not possible when hard, wear-resistant coatings of 1–10mm are used. Rickerby and his co-workers [12,14,15] have theoretically estimated the hardness of the coating by an analytical method (*the law of mixtures*) from the hardness value obtained under increasing loads.

In this respect, the newly developed *nano-indentation* technique [16,17] is considered to be more appropriate for thin coatings. This technique has been reported to be successful in determining the hardness, elastic modulus and creep behaviour of very thin coatings [18, 19]. However, the extent of time and cost involved in performing a test will prohibit its wide application.

11.4.2 Adhesion of coating

The single most important parameter in surface engineering is the *adhesion* between the coating and its substrate. There are many methods available for estimating the strength between the coating and its substrate, such as the *scratch test*, *laser spallation test*, *shear test*, *pull test*, *adhesive tape test* among many mechanical and non-mechanical tests [20]. Among all, the scratch test has been considered to be more appropriate and fairly accurate for hard coatings. The scratch test was originally proposed by Heavens [21], who used a smoothly rounded steel point. It was further developed by Benjamin and Weaver [22]. It should be mentioned here that there is no scientific basis behind the determination of the *critical load* at which the coating is considered to have failed in a scratch test. The critical load depends not only on the adhesive strength between the coating and substrate, but also on the hardness, roughness and thickness of the coating, the hardness of the substrate and the friction coefficient between the slider and the coating [23]. For example, Fig. 11.3 shows the variation of the critical load with roughness of the substrate (tool steel) for TiN coatings [24]. Nevertheless, the critical load does enable one to obtain a ranking. This diagnosis of the coating failure after scratch testing has to be validated by one or more of the following surface analytical tools:

- optical and/or scanning electron microscopy,
- surface topological analysis (3-D profilometry), or

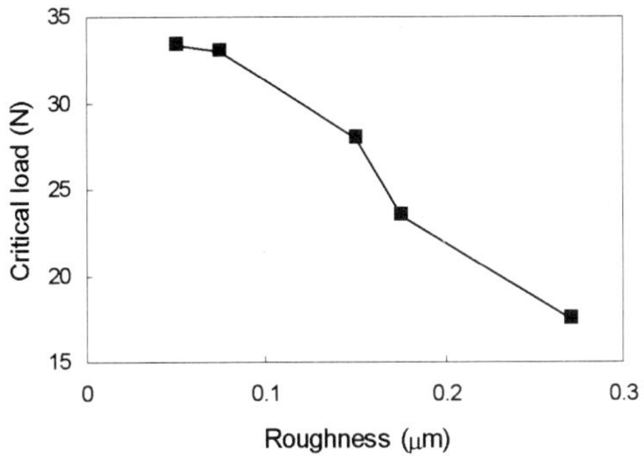

Figure 11.3 The variation of scratch adhesion of 3 μm thick TiN coating deposited on a tool steel substrate.

- chemical analysis by energy dispersive X-ray analysis/electron beam micro analysis [25].

11.5 Wear testing of coatings

As mentioned before, the best method for testing a coating for its wear behaviour is to put it into service. As this is expensive, time consuming and sometimes catastrophic, we have to have some testing facilities which are simple and easy to use and at the same time, relevant to the application of a particular component.

The wear test methodology should be chosen depending on the thickness of the coating. Thicker coatings should be treated as bulk monolithic materials whereas for thin coatings different criteria should be applied. In the following sections, we will discuss:

- selection of wear testing machine,
- selection of wear rate measurement method/criteria,
- methods of wear failure analysis/techniques, and
- comparison of wear data – laboratory to field tests.

11.5.1 Selection of wear testing machine

The American Society of Lubrication Engineers (ASLE) (presently the *Society of Tribologists Lubrication Engineers*, STLE) published a list of more than two hundred different types of wear testing machines in the mid-1970s [26]. These machines can be grouped as in Table 11.2. The *pin-on-disc* machine has been considered to be a de facto standard [27]. Other two popular machines are pin-on- ring and reciprocation-type rigs. For coatings, generally the flat pin is replaced by a ball or hemispherical pin which will facilitate the seating of the pin or ball

Table 11.2 The ASLE classification of wear testing machines

Class	Description	Typical device and title
A	Multiple sphere	Four-ball lubricant tester
B	Crossed cylinders	Crossed cylinder test apparatus
C1	Pin-on-flat (moving pin)	Pin fretting test
C2	Pin-on-flat (moving flat)	Kinetic boundary friction tester
C3	Pin-on-flat (multiple contact)	Oxide scale friction apparatus
D	Flat-on-flat	Stick-slip test apparatus
E	Rotating pins-on-disc (face loaded)	Friction and cold welding rig
F	Pin-on-rotating disc (face loaded)	Disc rider friction apparatus
G	Cylinder-on-cylinder (face loaded)	Solid lubricant test machine
H	Pin-on-rotating cylinder (edge loaded)	Pin-on-ring tester
I	Flat-on-rotating cylinder (edge board)	Dual rub shoe tester
J	Disc-on-disc (edge loaded)	Rolling disc machine
K	Multiple specimens	Cylinder-on-ball rolling fatigue test
L	Miscellaneous	Abrasive belt wear test machine

without much trouble on the counterface. These machines are generally used in sliding wear situations.

For *abrasive* wear testing, there are many types of machines available [28]. Some of them have been used for evaluating abrasive wear resistance coatings [29]. *Rabinowicz's* wear testing technique, using a lapping machine [30], was modified by Singer and his co-workers [31,32], which gives a depth resolution of 10 nm. Similar to depth profile of an element obtained by Auger analysis, this technique provides wear rate/resistance as a function of depth from the surface. In one case, the wear resistance profile was matched to N^+ concentration profile in an N^+ ion-implanted carbon steel [31].

There are few problems with very small abrasives especially with softer specimens (< 400 H_v). However this lapping method is suggested to be very useful in estimating the abrasive wear resistance of very thin layers such as the one obtained in ion-implanted materials and very thin PVD coatings. This method has been successfully used to evaluate the wear behaviour of ion-implanted 304 stainless steel, 52100 steel and titanium alloys, coatings produced by PVD and CVD methods and laser-alloy surface composites.

The selection of the wear testing machine depends on many factors such as the closeness of the service condition to the testing machine, availability, etc. In many instances, the availability of the wear testing machine determines the type of tests to be used. This is because a commercially available machine costs £7,500 –£75,000 or more, depending on the sophistication involved. Generally home-made ones cost a great deal less. The lack of funding for tribology studies forces researchers to use available testing machines even though they are partly inappropriate for the intended use.

Having made the decision on the type of testing rig, the next question is *which is the specimen – the pin or the disc?* Other related questions involve the type of counterface, levels of load and speed, type of lubrication (or dry), etc. These are important parameters because some studies have proved that there is a significant difference in the wear rates by reversing the pin and disk rôles [33], and by changing the counterface [34].

The type of testing machine or machines to be selected also depends on the modes of wear to be encountered in the actual application. For example, if the components in service failed by a mixed mechanism of adhesive and abrasive (*3-body low stress*) wear, it is prudent to select two machines which can simulate these mechanisms. Then there is a possibility that the test data obtained from these two sets will be relevant to the actual application. It would be more appropriate that, if possible, the amount of wear due to each mechanism should be determined and a rule of mixtures can then be developed, for example,

$$W_a = \sum f_n W_n \qquad [11.1]$$

where W_a = wear rate of the components in actual applications
f_n = volume fraction of wear due to wear mechanism, n
W_n = wear rate of the specimen in laboratory under wear mechanism.

If these wear mechanisms interact among themselves, then there is a need to consider *synergistic* effects. In such a synergistic situation, *the rule of mixture* may be inadequate. A term of interaction should be added to the above equation.

11.5.2 Wear rate measurement

The wear of the coating can be measured in many ways – volume or mass loss, dimensional change, etc. Depending on the density of the coating material, sometimes the weight loss method results could be misleading, especially when comparing two coatings with differing densities. In such situations, volume loss or dimensional change would be appropriate. Other wear measuring/monitoring techniques include visual or tactile inspection, profilometery, radiography, spectroscopy of wear particles and acoustic signal measurement.

These wear volumes can be expressed as wear rate which is wear volume/unit sliding distance or time. The thickness of the coating and sliding speed of this test are important here, as these two will determine the duration of the test. As mentioned in the *Metals Handbook*, the best test duration for a thin coating is until it wears out [35]. Thicker coatings can be tested for a longer time before the wearing surface reaches the coating/substrate interface. In many instances the arrival of the wearing interface to the substrate is hard to judge. However, in some cases, the change in colour of the coating, friction coefficient or noise could be an indicator that the coating is worn out.

Measurement of wear volume or change in dimensions should be continuously monitored or periodically measured. This will facilitate the wear rate measurement at all time intervals so that the history of the coating failure in wear testing can be known. Along with wear, friction coefficient is also generally recorded.

In general, wear by mass loss is measured using a sensitive balance (accuracy 0.1 mg). This determines the maximum weight of the specimen which can be weighed in the balance with the required accuracy (normally below 65g). The heavier the specimen the less accurate the reading. If heavier specimens are to be used, surface profilometry can be employed to measure the wear volume from the profile of the wear scar. There are contact and non-contact type surface profilometers with 2D or 3D mapping facility available commercially. The width or more generally, the wear depth of the wear scar is used as a criterion of wear.

11.5.3 Wear maps

Once wear volume or weight loss is calculated it is generally expressed as wear rate with respect to distance or time of sliding (mm^3/m or mm^3/s). Wear rates are plotted against applied load or sliding velocity. Often *specific wear rate* is mentioned which is the wear rate divided by the applied load (mm^3/mN). When specimens of varying size and shape are used by different researchers, it is difficult to compare the results. This deficiency has been addressed by Lim and Ashby [36] who have suggested the use of *normalized* parameters so that wear results from different sources can be compared. The normalized parameters for load, velocity and wear rate are dimensionless as shown below.

Normalized pressure, $$F_n = \frac{F}{AH}$$ [11.2]

Normalized velocity, $$V_n = \frac{V}{v}$$ [11.3]

Normalized wear rate, $$W_n = \frac{W}{A}$$ [11.4]

where F = applied load, N
A = area of contact, m^2
H = room temperature hardness, Pa
V = sliding velocity, m/s
v = thermal diffusivity, m/s
W = wear rate, m^3/m.

Wear mechanism maps are constructed with normalized pressure and normalized velocity as *x* and *y* axis whereas normalized wear rates are shown as contours. Following the wear mechanism map for steel [36], wear maps for other systems have been proposed, for example, aluminium alloys [37,38], ceramics [39], abradable coatings [40] and cutting tools [41]. These wear maps have been summarized recently as useful databases for wear applications [42].

11.5.4 Post-wear analysis

There is a gamut of analytical techniques available for examining the wear debris particles and the worn surfaces [43,44]. Among them, the most commonly used techniques are *scanning electron microscopy* (SEM) of wear particles and worn surfaces for characterising the morphology of debris and surface appearance of worn surfaces. SEM combined with analytical facilities would be very useful in characterising wear mechanisms. X-ray diffraction is used to determine the phases present in the debris and sometimes on the worn surface. The particle size distribution of the wear debris is often determined by a laser-based particle size analyser. Other widely used techniques are *X-ray photoelectron spectroscopy* (XPS), *Auger electron spectroscopy* (AES) and *transmission electron microscopy* (TEM).

11.5.5 Correlation between laboratory and field test data

As mentioned before, the wear data obtained in a laboratory wear testing machine does not represent the true performance in the actual field applications. This is because of differences in specimen geometry, size, type of environment, load and speed, to mention a few. Depending on the load or speed, the wear mechanism changes from one to another. These aspects have been discussed at length in the papers dealing with wear mechanism maps [36–42].

Wear mechanism maps are drawn using dimensions of pressure and velocity as axes with the wear rate contours (refer to *Section 11.5.3*). These maps can be used to predict the mechanisms of wear under certain conditions. These maps can also be applied to actual field conditions such as metal cutting – Fig. 11.4 [41].

11.6 Case studies

The following are a few examples of wear testing in an attempt to evaluate the wear behaviour of components or tools used in industry. Of particular interest is the comparison of selected parameters of wear testing in laboratory to those encountered in real-life situations.

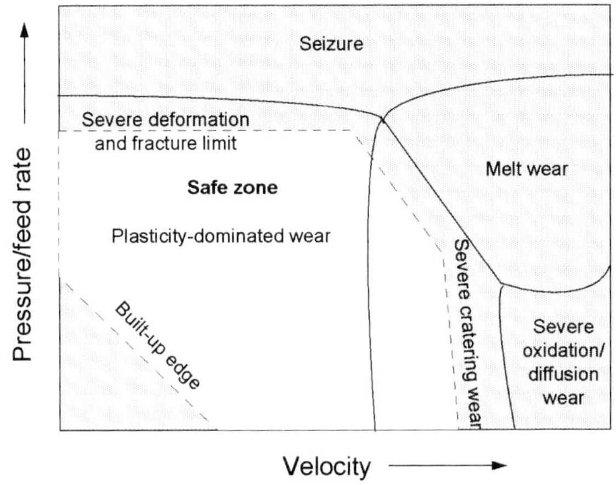

Figure 11.4 Application of wear map concepts to metal cutting situations [41]. It also shows safe régimes for cutting tools.

11.6.1 Wear of cutting tools

Direct and indirect measurements of various tool wear criteria have been reported [45]. One direct method of measuring tool wear is to stop the machining operation periodically and record the amount of wear (crater, depth, wear on the clearance face, etc.) either using optical or electron microscopes or through special instruments which measure the topographic features of the tools. Indirect measurements of wear of cutting tools include measurements of cutting force, cutting temperature, tool deflection, tool vibration, power input to the machine tool, autoradiography of wear particles and electrical resistance between the workpiece [46–49]. These indirect measurement techniques are vitally important as a new era of machining enters complete automation. However, it should be mentioned here that more work needs to be done in correlating such indirect measurements of tool wear to the actual wear.

The running of a machine test includes selection of various parameters involved in machining such as machine tool to be used, tool geometry, feed rate, the depth of cut, lubrication, workpiece material speed etc. Generally, progressive tool wear is measured for various speeds keeping all other parameters constant. A *Taylor* tool life plot of log tool life vs. log speed is established for various tools and thus the performances of tools is compared [46].

Mostly in metal cutting situations, the wear is best simulated using similar machine tools such as a lathe or radial drilling machine. Such tests are generally reproducing the service conditions and thus wear mechanisms. Tables 11.3 and 11.4 show that there are more similarities between the real-life and the laboratory situations [50–52]. Figure 11.5 shows typical results which compare the performance of uncoated and TiN-coated twist drills in drilling AISI 1045 steel [50,51]. The increase in the maximum thrust force recorded just prior to the failure

Table 11.3 Wear of drills

Parameter	Real-life situation	Laboratory
Geometry	Drilling machines	Radial drilling machine
Lubrication	Dry/coolant/cutting fluid	Dry
Speed	20–200 m/min	20 m/min
Feed rate	Varies	0.015–0.25 mm/rev
Temperature	Ambient	Ambient
Materials	Drills (HSS)/work materials	Drill (HSS)/1045 steel
Wear mechanism	Abrasion/adhesion/plastic deformation	Abrasion/adhesion/plastic deformation

Table 11.4 Wear of cutting tools

Parameter	Real-life situation	Laboratory
Geometry	Metal cutting using lathes	Facing using a lathe
Lubrication	Dry/coolant/cutting fluid	Dry
Speed	110-550 m/min (1200 m/min for ceramic tools)	85–942 m/min
Depth of cut	0.5–5 mm	0.3 mm
Feed rate	0.2–0.8 mm/rev	0.25 mm/rev
Temperature	Ambient	Ambient
Materials	Tool/work material	Tool/work material
Wear mechanism	Abrasive/adhesive/chemical	Abrasive/adhesive/chemical?

of the drill should be noted. These signatures can be used to predict the onset of the failure of drills and thus usefully incorporated in the tool maintenance programme. Figure 11.6 shows flank wear of uncoated and coated cutting tools used against AISI 1045 steel [52]. A typical failure is shown in Fig. 11.7 depicting flank and crater wear; the presence of work material on the cutting edge is also seen.

11.6.2 Wear of metal forming dies and tools

Dies and tools with varying sizes and shapes are used in the metal forming industry. For example, Fig. 11.8 shows a view of the front panel upper wipe die which produces flanges for automobiles. Generally the dies and tools used in metal forming operations experience elastic deformation whereas the workpieces undergo plastic deformation. This is different from the metal cutting situation where severe plastic deformation of the workpiece and the tool occur generating high temperatures at the tool/chip interface. This is one of the reasons why certain surface treatments (e.g. ion implantation) performed very well under metal forming operations but failed in metal cutting situations.

A scan through the literature indicates that any surface treatment or coating

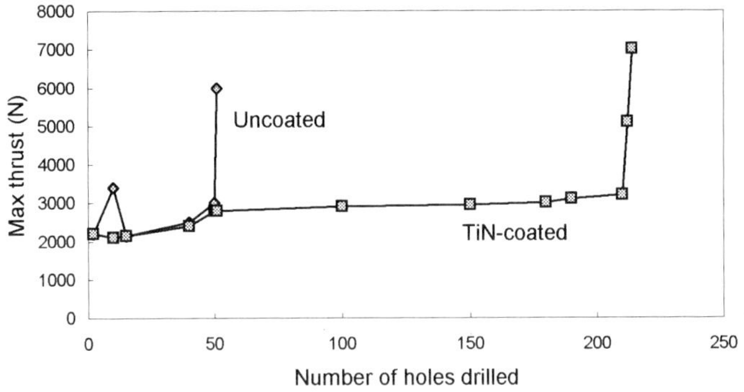

Figure 11.5 A comparison of uncoated and TiN-coated high speed steel drills [50,51].

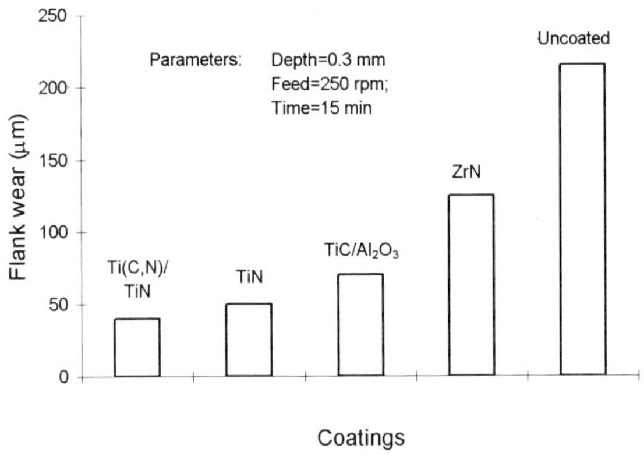

Figure 11.6 Flank wear of uncoated and coated carbide cutting tools (work material: AISI 1045 steel).

Table 11.5 Wear of metal forming dies

Parameter	Real-life situation	Laboratory
Geometry	Sheet metal sliding on tool	Pin-on-disc
Lubrication	Boundary lubrication	Dry
Speed	0.1 m/s	0.3 m/s
Load	Varies	2 kg
Temperature	Ambient	Ambient
Materials	Sheet metal/tool steel	Ceramic/tool steel
Wear mechanism	Adhesive/abrasive	Abrasive/adhesive

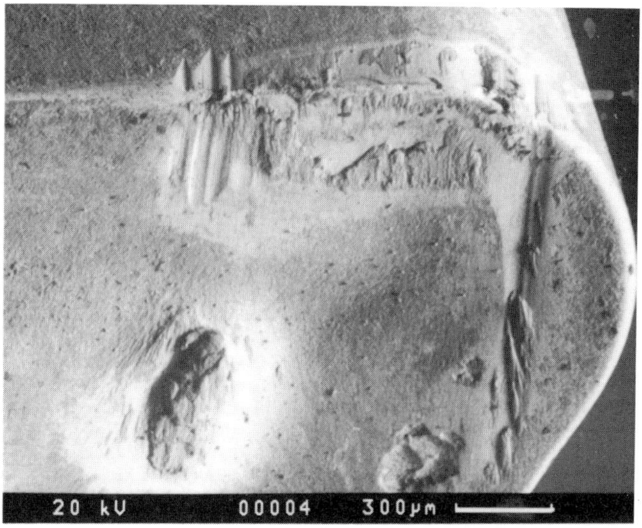

Figure 11.7 SEM micrograph of an uncoated carbide tool after cutting AISI 1045 steel for 15 min showing both flank and crater wear as well as some transferred work material (depth of cut = 0.3 mm; feed rate = 0.25 mm/rev; speed = 250 rpm).

Figure 11.8 A view of front panel upper wipe die which produces flanges for automobiles. The front inserts are vanadised whereas the rear ones are chromium plated (courtesy: General Motors – Holden, Salisbury, SA, Australia).

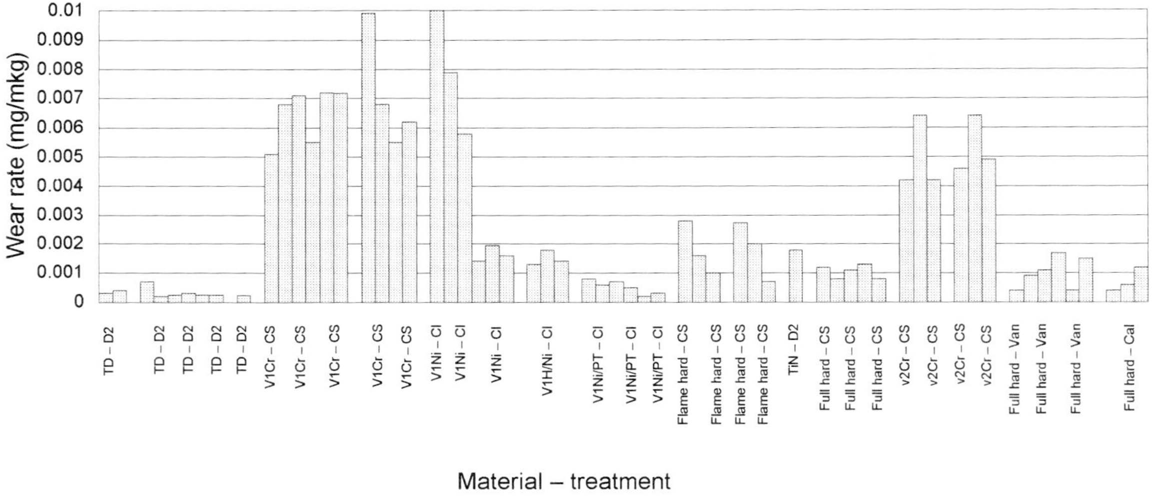

Figure 11.9 Wear rates of selected surface treatments/coatings on cast iron (CI), cast steel (CS) and tool steel (D2) substrates using a pin-on-disc wear tester. The counter surface (ball) was yttria-stabilised zirconia under a load of 2kgf and speed of 0.3 m/s.

applied to metal forming dies and tools has largely been tested under actual application conditions [53], though in some cases, simulation of such operations has been performed [54]. Wear in such dies and tools involves adhesion (metal pick-up), abrasion, fatigue (spalling) and corrosive wear (when operated in a corrosive environment). As mentioned previously, simultaneous occurrence of two or more such basic wear modes makes any simple laboratory test inappropriate. Also, as mentioned earlier, the *rule of mixtures* formula could be worth considering.

When the adhesion, hardness and other basic characterisation of a coating are evaluated, it is suggested that the coating be tested in actual metal forming operations. Periodic monitoring of wear of dies/tools or the workpiece by visual inspection for obvious pick-up, profilometry and microscopy should be carried out.

Table 11.5 shows various parameters which are considered to be important in simulating the wear conditions of metal forming dies in laboratory [55, 56]. Figure 11.9 shows the wear rates of die materials with different treatments tested in a pin-on-disc geometry. It should be borne in mind that the simulated conditions are not exactly reproducing the actual field conditions and therefore limit its application.

11.6.3 Wear of gears

Gears are used in many engineering applications and need frequent replacement because they fail generally due to wear and fatigue. Gears are tested by many methods including bending fatigue tests (using single tooth bending or reciprocating gears) for fatigue resistance and contact fatigue tests (using roll-on-roll rig, block-on-roll rig or reciprocating gears) for wear/contact life. The contact fatigue aspect is of interest now as it involves wear of gears by spalling which limits the useful

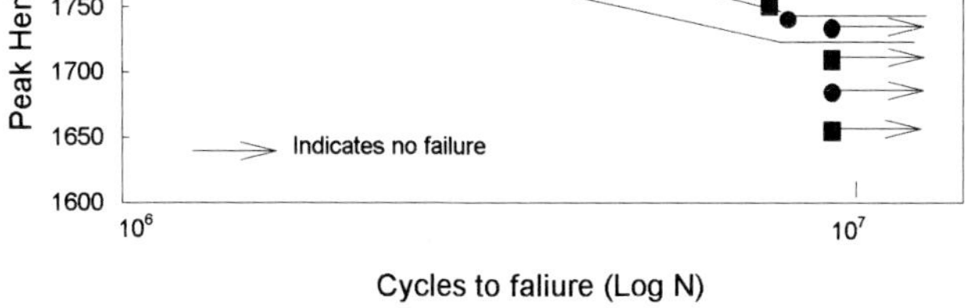

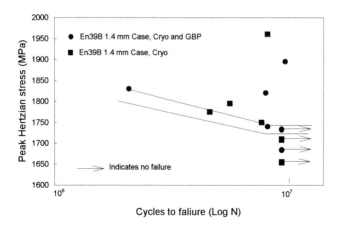

Figure 11.10 Contact fatigue performance of EN39B steel with different surface treatments after carburising: cryo = cryogenic treatment and GBP = glass bead peening.

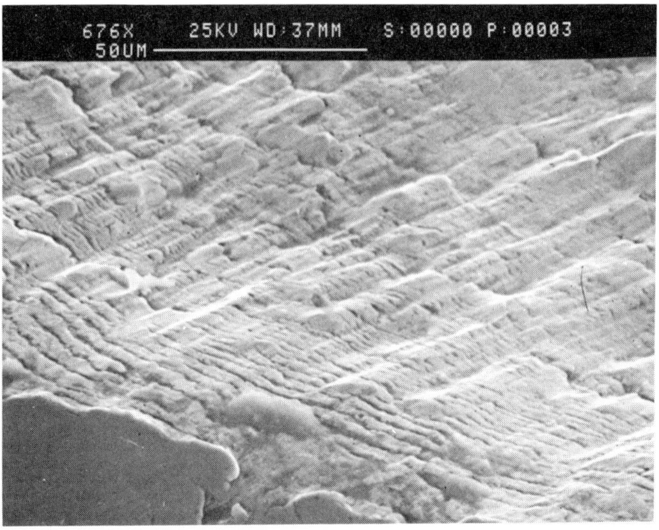

Figure 11.11 A typical contact failure in a carburised EN39B steel exhibiting fatigue striations in the crack surface.

Table 11.6 Wear of gears

Parameter	Real-life situation	Laboratory
Geometry	Gear	Contact fatigue (roll-on-roll)
Lubrication	Uneven; contaminated	Uniform; filtered
Speed	Varies	1500 rpm ϕ75 mm
Load	Varies	200–1200 kg (1200–2200 MPa Hertzian stress)
Slip	0–7%	3%
Temperature	80–110°C	80°C (hot oil)
Materials	Steel/steel	Same as in real life
Wear mechanism	Spalling, abrasion	Spalling

life of gears. Figure 11.10 shows the contact fatigue performance of EN39B steel with different treatments [57,58]. Spalling of the surface of the gears occurs due to rolling/sliding motion under applied load, as shown in Fig. 11.11. Table 11.6 shows how fatigue and wear of gears can be simulated in the laboratory using roll-on-roll tester [57,58].

Figure 11.12 A break-down roll for a heavy structural mill (courtesy: BHP Long Products Division, Whyalla, SA, Australia).

11.6.4 Wear of hot rolling mill rolls

Rolls used in hot rolling mills undergo extensive wear and need frequent replacement resulting in loss of production and expenses. A typical roll used in hot rolling mills is shown in Fig. 11.12. Many laboratory methods have been

Table 11.7 Wear of hot rolling mill rolls

Parameter	Real-life situation	Laboratory
Geometry	Rolling	Roll-on-roll
Lubrication	Water/water oil	Dry/water?
Speed	280 rpm (ϕ 1.4 m)	Roll 40–700 rpm (ϕ 40–70 mm) Steel 20–300 rpm (ϕ 80–100 mm)
Load	1200 ton (~1.2×10^7 N)	< 2000 N
Slip	0–100%	0–200%
Temperature	800–1200 °C	700–1100 °C
Materials	Cast iron to cermet-work material	Adamite/1045 steel
Wear mechanism	Abrasion, corrosion, adhesion, fatigue	Abrasion, corrosion

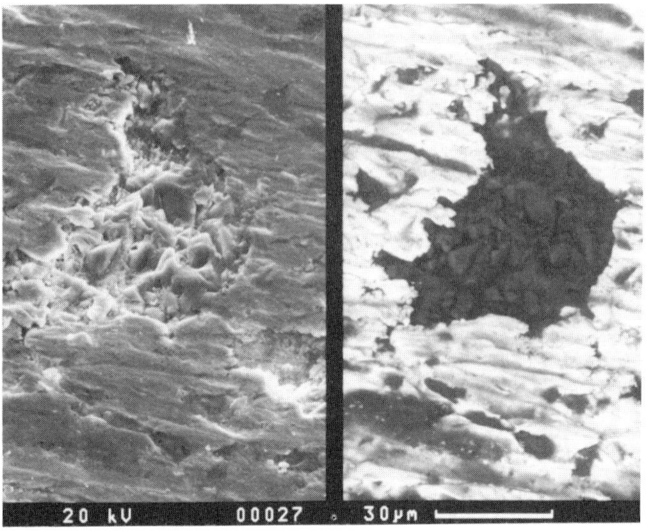

Figure 11.13 SEM of a steel ploughshare showing the entrapment of a hard particle from the soil in secondary and backscattered electron image modes: abrasive wear marks and plastic flow of steel over the silica-rich particle is evident.

suggested to simulate the wear behaviour of rolls in real-life situations such as roll-on-roll tester, block-on-ring tester and scaled down version of the actual rolling mill itself, in addition to simple "*pure*" single groove scratch testing at elevated temperatures.

In a recent project at the University of South Australia [59,60], two different test methods are used to evaluate roll materials for their wear behaviour, namely scratch abrasion tester and roll-on-roll rig. These tests are performed both at room and elevated temperatures. Table 11.7 shows a comparison of selected parameters in the simulated roll-on-roll testing and actual rolling mills. The level of the parameters are chosen in order to simulate the actual mechanisms of debris production, i.e. wear mechanisms.

11.6.5 Wear of tillage tools

The wear of ground-engaging tillage tools is a major problem to many Australian farmers. These tools are subject to very high rates of wear, depending on the soil type and ploughing conditions employed. The wear problems of these agricultural tools have been studied for many years in order to improve their lifetimes. Many different testing methods are used to test various materials to rank them in order of performance. Assuming abrasion is the main cause of wear, either rubber wheel

Table 11.8 Wear of tillage tools

Parameter	Real-life situation	Laboratory
Geometry	Plough shares – tool/soil	Rubber wheel abrasion tester
Lubrication	Dry/natural soil moisture	Dry/wet
Speed	4–12 km/h	0.8 m/s
Load	Up to 4000 N	130 N
Temperature	Ambient	Ambient
Materials	K 1073 or Ni-hard cast iron/soil	Steel ploughshare/alumina or garnet
Wear mechanism	Severe abrasion	Abrasion

abrasion testers or pin-on-disc testers were used. More close to reality are the simulated paddle-type testers where the specimens are immersed in a controlled soil and rotated.

These simulated tests will have no correlation with the real-life problem unless the wear mechanisms are reproduced in laboratory. Figure 11.13 is a scanning electron micrograph showing both secondary and backscattered electron images of a particle picked up by the steel plough from the soil and subsequent plastic deformation [61]. When tested in the laboratory with sandy soils, such pick-ups are not observed. It is obvious that bigger stones and rocks – bigger than the size of the tool often found in agricultural fields – cannot be included in any laboratory scale testing rig. Alternatively, it is suggested to use two more testing methods in order to simulate different wear mechanisms, once they are identified. Table 11.8 shows how a rubber wheel abrasion tester can simulate the abrasive component of tillage tool wear [62].

11.7 Conclusions and recommendations

The wear testing of coatings and surface-modified materials is similar to those used for bulk materials except the fact that the sensitivity of the measurement of wear rate should be higher in the former. The large variations in the results of these wear rates of coatings may be due to overlooking of the thickness and the variation of properties from microstructure, composition and/or from the surface towards the bulk.

A *rule of mixture* can be applied to predict the performance of the coating in the actual situation from the wear rate data obtained in the laboratory machines simulating various wear mechanisms. Wear mechanism maps with parameters, which are relevant to the real-life situations, should be developed to enable better correlation between laboratory and field data.

Acknowledgements

The author would like to acknowledge Prof Ken Strafford, the Head of Gatrell School, for his encouragement and support during the preparation of this article. Thanks are also due to Messrs J Cavallaro, S Ferguson, M Gopinath, J Lundy, W McMillan, M McPhee and S Spuzic for their timely assistance regarding the details of various research projects.

References

1. Jost, H P, (Tribology) Education and Research, The Jost Report, London, H. M. Stationery Office, 1966.
2. Czichos, H, *Wear* **41**, p45, 1977.
3. Burwell, J T and Strange C. D., *J. Appl. Phys.*, **23**, p18, 1952.
4. Lancaster, J K, *Wear*, **141**, p478, 1991.
5. Codet, M, *Wear*, **100**, p437, 1984.
6. Bell, T, *Metals Mater.*, **7**, 478, 1991.
7. Bayer, R G (ed), The Selection and Use of Wear Tests for Metals ASTM Special Tech. Publ. p615,1976.
8. Wear Tests for Plastics; Selection and Use, ASTM Special Tech. Publ. p701, 1979.
9. Selection of Wear Tests for Coatings, ASTM, Special Tech Publ. p769 1982.
10. Selection and Use of Wear Tests for Ceramics, ASTM Special Tech. Publ. p1010, 1979.
11. Uetz, H, Sommer, K and Khosrawi, M A, *VDI Berichte*, **354**, 107, 1979.
12. Burnett, P J and Rickerby, D S, *Thin Solid Films*, **148**, 41, 1987.
13. Buckle, H, The Science of Hardness Testing and its Research Applications, (ed.) Westbrook J. W. and Conrad H., ASM, OH (1973) p453.
14. Bull, S J and Rickerby, D S, *Surf. Coat. Technol.*, **42**, p149, 1990.
15. Burnett, P J and Rickerby, D S, *Thin Solid Films*, **148**, p51, 1987.
16. Pethica, J B, Hutchings, R and Oliver, W C, *Philos. Mag.*, **48**, p593, 1983.
17. Field, J S, *Surface Coat. Technol.*, **36**, p817, 1988.
18. Pethica, J B, Ion Implantation into Metals, (ed) Ashworth, V Grant, W A and procter, R P M, Pergamon, Oxford, p147, 1982.
19. Bunshah, R F, *Thin Solid Films*, **107**, p21, 1983.
20. Rickerby, D S, *Surface Coat. Technol.*, **36**, p541, 1988.
21. Heavens, O J, *Phys. Rad.*, **11**,355, 1950.
22. Benjamin, P and Weaver C, Proc. Roy. Soc., A254, 1960.
23. Valli, J and Makela, V, *Wear*, **155**,1987.
24. Subramanian, C, Strafford, K N, Wilks, T P, Ward, L P and McMillan, W, *Surface Coat. Technol.*, **62**, p529, 1993.

25. Von Stebut, J, in Plasma Surface Eng. **2**. (ed). Broszeit, E, and Munz, W D, Oochner, H, Rie, K T and Wolf, G K DGM Germany, p1215, 1989.
26. ASLE Friction and Wear Devices, 2nd ed., 1976.
27. Eyre, T S and Davies, F A, in Surface Stability, (ed) Rhys-Jones, T. N., The Inst. of Metals, p186, 1989.
28. Blickensderf, R and Laird II, G J, *Testing and Evaluation*, **16**, p516, 1988.
29. Bull, S J, Rickerby, D S, Robertson, T and Hendry, A., *Surf. Coat. Technol.*, **36**, p743, 1988.
30. Rabinowicz, E, *Lubrication Engineer*, **33**, 7, p378, 1977.
31. Singer, I L, Bolster, R N and Carosella, C A, *Thin Solid Films*, **73**, p283, 1980.
32. Bolster, R N and Singer I L, ASLE Trans., **24**, p526, 1981.
33. Rice, S L, Wayne, S F and Nowotny, H, *Wear*, **88**, 85, 1983.
34. Subramanian, C, *Scripta Metall. Mater.*, **25**, 1369, 1991.
35. ASM Metals Handbook, 9th edition, Vol. 8, Mechanical Testing, 1985 601.
36. Lim, S C and Ashby, M F, *Acta Metall.*, **35**, p1, 1987.
37. Antoniou, R and Subramanian, C, *Scripta Metall.*, **22**, p809, 1988.
38. Liu, Y, Asthana, R and Rohatgi, P K., *J Mater. Sci.*, **26**, p99, 1991 .
39. Hsu, S M, Wang, Y S and Munro, R G, *Wear*, **134**, p1, 1989.
40. Borel, M O, Nicoll, A R, Schlapfer, H W and Schmid, R K, *Surface Coat. Technol.*, **39/40**, 117, 1989.
41. Kendal, L A, in Metals Handbook, 9th ed., vol 16, Machining, p37, 1989.
42. Lim, S. C., Liu, Y. and Lee, S. H., J. Inst. Engineers, Singapore, **31**, p51 1991 .
43. Weber, E R, Auger Electron Spectroscopy for Thin Film Analysis, *Res. Dev. Magazine*, **23**, 10, p22, October 1972.
44. Buckley, D H, Surface Effects in Adhesion, Friction, Wear and Lubrication, Tribology Series 5, Elseveir 1981.
45. Tipnis, V A, in Wear Control Handbook, (ed.) Peterson, M B and Winer, W O, ASME, NY, p891, 1980.
46. Trent, E M, Metal Cutting, 2nd ed., Butterworths, 1984.
47. Shaw, M C, Metal Cutting Principles, 3rd ed., MIT Press., Mass., p891,

Chapter 12

The use of organic and inorganic coatings to reduce friction, wear and erosion

Dr N R Whitehouse – Paint Research Association

12.1 Introduction

We are all familiar with the use of oils and greases to lubricate mechanical parts and reduce wear. These products are being improved constantly in an effort to cope with an increasing range of environments. There are still a number of service conditions, however, in which oils and greases do not perform satisfactorily.

Such lubricants may lose their effectiveness in equipment operating at high temperatures and in machinery operating at high speeds or under heavy loads. This occurs because high interfacial temperatures are generated by friction.

Wear can be minimized if self-lubricating coatings or treatments are applied to the moving parts. Hard coatings, applied by a range of application techniques, have been developed for dry or marginally lubricated conditions, in particular high temperature environments.

In this Chapter, the most common, bonded solid film lubricants will be described and some typical applications will be outlined. The use of organic coatings to reduce erosion will also be considered briefly.

12.2 Bonded solid film lubricants

Bonded solid film lubricants are unique products. They utilize the special properties of a variety of lubricating solid materials, in coating form. In simple terms, they may be thought of as '*slippery paints*', though this definition is not exact and does not describe adequately the range of products which have been developed and used successfully.

A solid lubricant may be defined as a material in the solid phase which leads to an improvement in sliding conditions when interposed between two sliding surfaces. It may achieve this by :

- reducing friction,
- reducing wear,

- reducing stick-slip, or
- preventing adhesion.

Although friction with dry lubrication may be much higher than friction with fluid lubricants, dry lubricants remain effective under a number of extreme conditions where fluid lubricants often fail. Examples are high temperature environments and inaccessible surfaces which cannot be re-greased.

Bonding methods include resin bonding, sputtering, ion deposition and mechanical impingement. Commercially, however, almost all of the materials in use today are *resin bonded.*

12.3 The formulation of bonded solid film lubricants

Typical formulations consist of a *lubricating pigment*, a *binder* (film-former), *solvents*, and *additives* (to improve application and corrosion resistance, for example). Each of these components will be considered in turn.

12.3.1 Pigments

In conventional coatings, the primary function of the pigment is to confer colour and opacity. In *bonded solid film lubricants*, however, the function of the primary pigment is to lubricate. The most commonly used materials are :

- molybdenum disulphide,
- graphite,
- polytetrafluoroethylene (PTFE).

Less commonly used materials include tungsten disulphide, boron nitride, fluorinated ethylene propylene and perfluoroalkoxy copolymers.

The choice of pigment for a particular application is determined by the performance required from the *bonded solid film lubricant*. Factors which are usually considered include :

- coefficient of friction,
- load carrying capacity,
- corrosion resistance, and
- electrical conductivity.

The environment in which the *bonded solid film lubricant* will have to perform, however, also needs to be considered and, in this respect, the most important factors are :

- temperature,
- pressure,
- humidity, and
- oxygen content.

Each solid lubricating pigment has its strengths and weaknesses. Compromises often have to be made. The performance characteristics of molybdenum disulphide, graphite, and PTFE are as follows :

Molybdenum disulphide. This compound has a high load-carrying capability with a correspondingly low coefficient of friction. The coefficient of friction decreases with increasing load, is almost independent of temperature at normal humidities, falls with temperature at lower humidities, and falls with rubbing time. The sulphur atoms have a natural affinity for most metal surfaces, especially surfaces which are clean and free from oxidation products. The upper working temperature limit for *bonded solid film lubricants* containing molybdenum disulphide is 400°C, in practical terms, because the compound begins to decompose (in an oxidative atmosphere) at temperatures above this figure.

Graphite. This material is used widely as a solid lubricant. Adsorbed vapours (mainly moisture) are needed to maintain good lubricating properties and so its use in dry environments (or low atmospheric pressures) is limited. Graphite has a high temperature capability in an oxidative environment, but tends to promote galvanic corrosion and will not function in high vacuum. The upper working temperature limit for *bonded solid film lubricants* containing graphite is around 450°C.

Polytetrafluoroethylene (PTFE). PTFE is the principal compound in a group of fluorocarbon polymers. It is inert to many wet and aggressive environments and provides dry film lubrication together with mechanical and electrical stability over a wide temperature range (–80°C to 250°C). PTFE is also resistant to most organic solvents. In the solid form (unbonded), it has poor mechanical strength, poor creep, and poor thermal conductivity. The coefficient of friction of PTFE falls with load and increases with speed. Because PTFE is inert and has very low adhesion, it is very difficult to apply to surfaces as a single material. Dispersed in a suitable binder, this poor adhesion can be overcome and bonded PTFE-containing lubricant coatings are the norm. PTFE's good frictional behaviour makes it one of the most useful pigments for bonded coatings.

Bonded solid film lubricants may contain more than one lubricating pigment. Graphite and molybdenum disulphide are often used in combination, for example, to introduce into a formulation the benefits of each material.

12.3.2 Binders

A wide range of high molecular weight film-forming materials (resins) is known to the coatings industry. Several generic types have been used as binders for conventional decorative and other high performance coatings and several of these film-formers have also been used as *binders* or bonding agents for solid film lubricants. The purpose of the binder in a solid film lubricant is to hold the functional pigment on the surface and so the binder itself need not have lubricating properties. Binders may be organic, inorganic, metallic or ceramic. Combinations of these types are also possible.

The *organic bonding agents* which have been looked at divide into three broad categories :

- ambient drying/curing types e.g. alkyds, acrylics and vinyls;
- chemically-cured types e.g. epoxies and urethanes;
- thermosetting types e.g. phenolics and silicones.

The *inorganic bonding* agents include: Silicates, phosphates and ceramics.

Selection is determined by the desired physical properties and the service environment, though solvent resistance, corrosion resistance and curing temperature are all equally important.

The significant properties of the organic bonding agents, which have been used in solid film lubricant coatings, are as follows:

Alkyds. These resins contain drying oils (linseed oil is an example); they cure (cross-link) by an oxidative mechanism. As a class of binder, they:

- are easy to apply,
- are tolerant of poor surface preparation,
- have poor water resistance, and
- have poor alkali resistance.

Alkyd resins are often modified with other resins to improve their water and alkali resistance. *Styrenated alkyds*, for example, are faster drying and have better chemical resistance.

Vinyls and acrylics. Vinyl and acrylic binders dry solely by evaporation of solvent. They are high molecular weight film-forming materials which are usually plasticized. Vinyls and acrylics:

- are quick drying,
- have good chemical resistance,
- have poor heat resistance,
- have poor resistance to strong solvents, and
- need to be applied to well prepared surfaces.

Epoxies and polyurethanes. These binders cure by chemical reaction. A base is mixed with an appropriate amount of curing agent immediately before application and the curing mechanism then begins. Two-pack epoxies and polyurethanes :

- have good water resistance,
- have good chemical resistance,
- have good hardness and abrasion resistance, but
- need to be applied to well prepared surfaces.

Phenolics and silicones. These binders need to be heated after application to develop their full properties. Phenolics and silicones:

- have excellent heat resistance,
- have good water resistance, and
- have good chemical resistance.

In general, the *organic air drying resins* are less expensive to apply, are amenable to field application and do not affect metallurgical properties. Their durability, wear life and solvent resistance, however, are not good. The *thermoset resins*, by contrast have good durability, solvent resistance and wear life. Only the *inorganic binders* have good high temperature stability. On the minus side, however, they are often difficult to apply.

12.4 Pigment/binder ratio

An important factor in the formulation of solid film lubricant coatings is the amount of lubricating pigment-to-binder. This is influenced by performance requirements and may be varied within a working range. Beyond a ratio, expressed in paint technology as the *critical pigment volume concentration* (CPVC), pigment particles are under-bound and the resulting coating is mechanically weak.

Low pigment-to-binder ratio formulations have higher coefficients of friction, appear glossy, have good corrosion resistance and tend to be durable. High pigment-to-binder ratio formulations, by contrast, have a low coefficient of friction, poor corrosion resistance and poor cohesive strength.

12.5 Solvents

The *solvents* used in *bonded solid film lubricant coatings* are dictated by the chemical nature of the binder in the first instance. Mixtures of solvents are used commonly in order to achieve optimum application qualities and acceptable drying times.

12.6 Additives

Additives (minor components) of many different types are added to paint formulations to optimize specific properties, such as flow, adhesion to substrate, and stability of a formulation. Additives used in *bonded solid film lubricant* coatings are similar to those used in conventional paints.

12.7 Composite coatings

Bonded solid film lubricants, in which the binder is organic, may soften under heat or be degraded by aggressive chemicals. For this reason, lubricating coatings have also been developed in which the binder is inorganic (most commonly metallic or ceramic). The main functional pigment used is PTFE as it is available in fine particle sizes.

Composite coatings of this type are applied either by metal spraying techniques or by plating and anodising processes. Stainless steel/PTFE, hard chromium plating/PTFE and electroless nickel/PTFE are the most common types commercially. Composite coatings have extended the usefulness of PTFE as a dry film lubricant to service conditions where organically-bonded types would not survive for an acceptable period of time.

12.8 Application of bonded solid film lubricants

Good *surface preparation* is essential and either degreasing and blast cleaning, or degreasing and phosphating are the preferred methods.

Resin-bonded solid film lubricants can be applied by conventional paint application methods such as spraying or dip-spin coating. Film thickness is important and must be controlled very carefully to give good uniformity. Typical thicknesses range from 5 to 15 μm though thicker coatings (25 to 50 μm) are usually needed if good corrosion resistance is also required. These are thin films compared with decorative coatings. Thin films ensure a uniform distribution of lubricating pigment throughout the coating and prevent a binder-rich layer on the surface. Bonded solid film lubricants are soft in comparison with the substrate. If applied too thickly, they have poor impact strength.

Selection of the method of application is governed by the number of parts to be coated, their size, the film thickness required, and the tolerances which can be accepted in production.

In order to achieve optimum coating performance it is necessary to *stove* many coatings after application. This *cross-links* the binder and leads to tougher coatings. In the case of a coating based on molybdenum disulphide, for example, typical temperatures range from 90°C to 200°C, depending on the binder type.

12.9 Typical applications

Resin-bonded dry lubricant coatings have been used successfully in many situations. Graphite in resin coatings has been used, for example, to coat piston skirts, to provide release for gaskets, and to modify the friction characteristics of brake linings.

Resin-bonded molybdenum disulphide coatings have been used for extreme pressure lubrication in gearboxes and engine valves.

PTFE-based products have been used for rubber seals and shock absorbers, windscreen wipers and washers, door hinges and sliding roof mechanisms. Threaded fasteners, for the automotive industry, are a large market for resin-bonded PTFE coatings. Resistance to corrosion is a prime requirement for this application and thin coatings have been developed not only to lubricate but also to protect.

For severe applications, such as marine environments, the best properties of resin-bonded PTFE coatings and zinc or cadmium plating have sometimes been combined for maximum performance. The use of cadmium in this way is likely to decline very rapidly, however, as environmental legislation tightens.

12.10 Erosion resistant coatings

One specialized area of wear resistant coatings is the prevention of erosion of surfaces by wet, dusty environments. Most of the development work in this area has been directed towards military applications and, in particular, aircraft radomes. Elastomeric coatings have been found to afford the best long-term protection. Neoprene was used initially but coatings based on this material were difficult to apply and weathered badly. They were also limited to service temperatures of 90°C or below.

Neoprene-based coatings have been superseded by poylurethane and fluorocarbon products, particularly vinylidene fluoride-hexafluoropropylene copolymers. The latest coatings can be applied in less than three hours to a final thickness of 0.3mm using conventional spraying equipment. They cure at room temperature and maintain erosion resistance, transmission and antistatic properties in service for three years or longer. Most recently, conductive fibres have been incorporated into topcoats as a replacement for carbon black. As a result, coloured coatings with improved camouflage properties can now be formulated.

References

The literature on this subject is fragmented and review articles are few. Papers by Bhushan and Gresham (and references therein) give additional information.

Bhushan B, *Metal Finishing*, 6,**78**, 5, p71, 1980

Gresham R M, *Metal Finishing*, 6, **86**, 3, p113, 1988

Chapter 13

Surface and interface analysis

A J Swift – CSMA Ltd

13.1 Introduction – Surface analysis in surface engineering

Surface analysis measures the composition and chemical/molecular state of the outermost atomic layers of all types of solid (and some liquid) materials directly. It is one of the fastest developing branches of analytical chemistry in material science and requires the use of complex and advanced instrumentation. It is nevertheless of great value to the surface engineer since application of the appropriate surface analysis technique(s) enables full chemical characterisation of the system whether in failure investigation or product development for example of a coating, an organic or inorganic system, electrical conductor or insulator.

Applications in catalysis stimulated the earliest developments of surface analysis in the early 1960s, principally due to the sensitivity of techniques such as *electron spectroscopy for chemical analysis* (ESCA, also known as *X-ray photoelectron spectroscopy*, XPS, see glossary) to the oxidation state of transition row metals and the crystal structure sensitivity of techniques such as *low energy electron diffraction* (LEED). However, materials scientists were quick to benefit from the emerging capability for true *surface* analysis with high chemical state sensitivity. Progress was made in the field of corrosion science in particular, where pioneering workers revealed new insight into corrosion processes on metal surfaces using ESCA and capitalising on the spatial resolution capability of *auger electron spectroscopy* (AES); again both in model system studies as well as retrospective failure analysis [1].

High spatial resolution in *auger* and *secondary ion mass spectrometry* (SIMS) and development of well characterized sputter depth profiling analysis using both techniques widened further the rôle of surface analysis throughout the late 1970s and 1980s into tribology, surface treatment and coating evaluation. The range of materials examined also widened to incorporate metal alloys and oxides, ceramics, semiconductors and polymers.

In recent years the introduction of convenient monochromatized X-ray sources and high performance time of flight mass spectrometers in SIMS has enabled organic materials to be probed with greater molecular sensitivity so that the performance of polymer coatings, adhesives/composites, organic films of corrosion inhibitors or lubricants, can all be characterized. Combined with improvements in spatial resolution in the imaging performance of both SIMS and XPS techniques, applications of surface analysis in *surface engineering* and materials science today are at a maximum. This is so at the same time that the technological requirements of modern industrial coatings and treatments are also rising sharply such that advanced *surface engineering* and surface analysis are required in the development of everyday items ranging from disposable soft packaging materials to contact lenses.

Chemical analysis of the surface layers of materials is a difficult task with unique considerations compared to other methods of analytical chemistry or materials microanalysis. Conventional practices relating to sample selection, handling the data interpretation etc., do not generally apply to the case of surface analysis as a consequence of the special behaviour of the surface region of a material.

This Chapter describes the essential features of the most commonly used surface analysis techniques in *surface engineering*, the mechanism of signal generation, its interpretation and the relative merits and limitations of these methods. Where relevant, these aspects are illustrated by reference to specific examples taken from industrial studies.

13.1.1 What is the surface and why is it important?

In surface analysis terms, the *surface* is defined as the outermost few atomic layers (0.1–10 nm) of the material. It will be shown in this Chapter that, without special sample preparation, for many modern surface analysis techniques this is the *maximum* sampling depth and that a great deal of chemical information can be acquired specifically from this region.

Beyond this region is the *near-surface*; ranging from 10nm to 1μm. Some surface analysis techniques examine this region effectively such as *depth profiling* SIMS/Auger. This is also the analysis régime of *scanning electron microscopy* (SEM) and SEM related techniques such as *energy dispersive X-ray analysis* (EDXA or EDAX) or *electron probe microanalysis* (EPMA). These latter methods often provide the most convenient first line approach for the investigation of near-surface or interface problems. SEM combined with EDAX provides a picture of the morphology and a semi-quantified analysis of the elements from a bell-shaped analysis volume in the uppermost ~1–3μm of a sample at high spatial resolution and high sensitivity (up to 0.1wt%). The elemental distribution over this volume from micron scale areas can also be mapped. Using modern instruments the full range of image processing and statistical analysis software is available for subsequent data treatment of physical and elemental images.

The application of SEM in its various forms and elemental analysis in *surface engineering* is immense. A great many volumes have been written on their application in materials science and failure analysis. Certain bulk analysis methods have also been adapted for investigation of the near-surface; for example infrared spectroscopy using *attenuated total reflection* (ATR–IR). However, instead of attempting to précis these methods, the reader is directed to other introductory texts [2,3] so that this chapter can focus on the instrumentation, information content and challenges of the main techniques for *surface* analysis in *surface engineering*.

The chemical properties of the surface or at key interface regions can often be critical in determining a material/product performance. In many cases the surface chemistry of a material is quite different to its bulk and, since it is this region which meets the outside world, understanding the chemical properties of the surface of the material can be just as important as the bulk analysis. This is the case for all types of solid materials from polymers to metals to glasses. Properties of adhesion, colour, corrosion resistance, biocompatibility and lubrication for example, can all be controlled in the surface region. Precise measurement of these properties is clearly of importance for a wide range of industrial products.

Several books have been written on specific techniques of surface analysis although these are mostly for practising surface scientists and go into some detail (see for example [1] for ESCA and [4] for SIMS). For a more general and comprehensive picture the reader is referred to recent publications [5–8].

13.1.2 General considerations

The principal techniques used for analysing the surface chemistry are based upon electron spectroscopy and mass spectrometry. Vibrational spectroscopy and scattering techniques are also available for surface investigation although these latter methods tend to be restricted in the breadth of their areas of application.

In this Chapter, ESCA, Auger and SIMS will be described in some detail since these are the most widely applied methods of surface analysis. The principle of signal generation and interpretation for each technique will be described and the nature of the information which can be acquired using these methods is illustrated by reference to a range of examples from a typical industrial laboratory. General criteria for a range of modern techniques are summarized in Table 13.1.

13.2 Electron spectroscopy

XPS and AES are linked as techniques since they are both dependent on the analysis of low energy electrons (in the range of 10–3000 eV) ejected from the surface. As a result these techniques yield essentially similar information (see Table 13.1) and both can use the same electron spectroscopy instrumentation.

13.2.1 X-ray photoelectron spectroscopy (XPS or ESCA)

In XPS the sample is irradiated with a beam of monochromatic soft X-rays.

Table 13.1 A summary of performance criteria of the most commonly used techniques for applied surface analysis

Technique	XPS	AES	LIMA	Static SIMS	Dynamic SIMS
Type of sample	Solid and low vapour pressure liquids	Conduction solid	Solid and low vapour pressure liquids	Solid and low vapour pressure liquids	Solid and low vapour pressure liquids
Species in/out	X-rays/electrons	Electrons/ electrons	Laser/ions	Ions or atoms/ions	Ions/ions
Sampling depth	10–50Å	5–30Å	0.5–2μm	5–10Å	5Å–5μm
Analysis area	40μm–5mm diameter	< 2000Å several mm	~ 1μm diameter	2500Å–5mm diameter	2500Å–5mm diameter
Imaging	Yes < 10μm resolution*	Yes, 0.2μm resolution	No	Yes, 1μm resolution	Yes, 1μm resolution
Elements detected	All except H and HE	All except H and HE	All	All	All
Information gained	Elemental, chemical	Elemental, some chemical	Elemental, some moleculars	Elemental and molecular/ polymer structure	Elemental and simple molecular
Structural information	Fair	Poor	Poor	Good	Fair
Quantification	Good	Good/fair	Poor	Semi	Good with standards
Sensitivity	100 ppm	500 ppm	1 ppm	ppm–ppb	ppm–ppb
Destruction	Low	Medium	High	Low	High

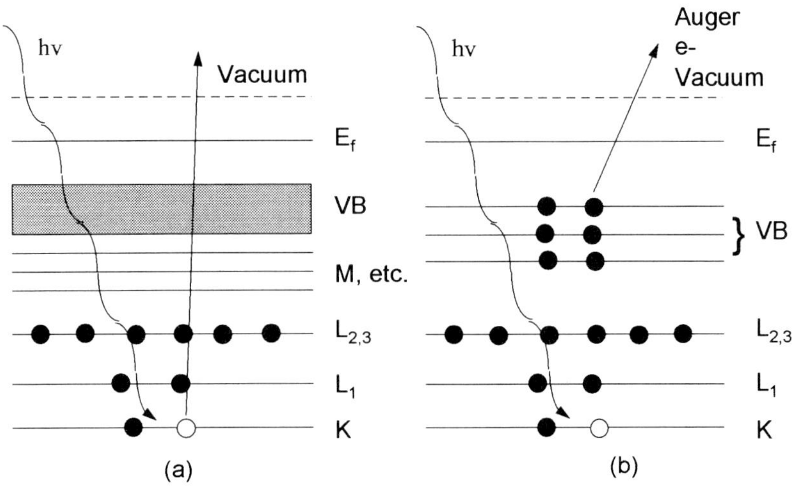

Figure 13.1 (a) and (b); Schematic diagrams of the XPS and Auger electron emission process. Where, Ef refers to the Fermi edge, VB the valence band and M, L, K refer to the spectroscopic notation of atomic energy levels.

Photoelectron emission results from the atoms in the specimen and the kinetic energy distribution of the ejected photoelectrons is measured directly in the electron spectrometer. A schematic diagram of the X-ray photoemission process is shown in Fig. 13.1 (a).

Each atom, except hydrogen, possesses core level electrons which are not directly involved with chemical bonding but whose energies are influenced slightly by the chemical environment of the atom. The binding energy (E_B) of each core level electron (approximately its ionisation energy) is characteristic of the atom and specific orbital to which it belongs. Core level binding energies of the elements in the periodic table are available in any handbook or practical text on ESCA, e.g. [1].

When the sample is irradiated with a beam of low energy monochromatic X-rays of photon energy *hv*, *Einstein's equation* states that for the conservation of energy:

$$E_K = hv - E_B - \O \qquad [13.1]$$

where E_K is the measured energy of the ejected electrons (eV)
$\O$ is the work function of the spectrometer
hv is the photon energy: $Mgk_\alpha = 1253.6$ eV, $Alk_\alpha = 1486.6$ eV

As the terms hv and $\O$ remain constant, the measured kinetic energy of a core level photoelectron peak can be related directly to its characteristic binding energy. Consequently, the photoelectron spectrum reflects the shell-like nature of the quantized electronic structure.

Figure 13.2 shows a typical ESCA spectrum recorded from a corroded steel surface. The elemental composition of the surface is calculated from peak area measurement of the peaks in the survey spectrum which have been normalized and corrected using a standard set of sensitivity factors. The spectrum shows that carbon, oxygen and iron are the principal elements in the surface, and that low levels of other alloy metals (copper, tin, molybdenum and chromium) are present along with some chlorine which is indicative of the corroding brine medium.

13.2.2 Auger electron spectroscopy (AES)

After photoemission of a core electron, the ion is left in an excited state and must decay back to the ground state for the conservation of energy. Consequently, energy is released when a lower energy electron drops back into the core hole. This energy can escape as an X-ray photon (X-ray fluorescence) or it can eject a third electron as shown in Fig. 13.1 (b). This is an *Auger electron.*

The Auger process is a relaxation event, where an excited state atom spontaneously ejects an electron with a characteristic *kinetic energy EK. The* only requirement for the ejection of an Auger electron is the formation of an

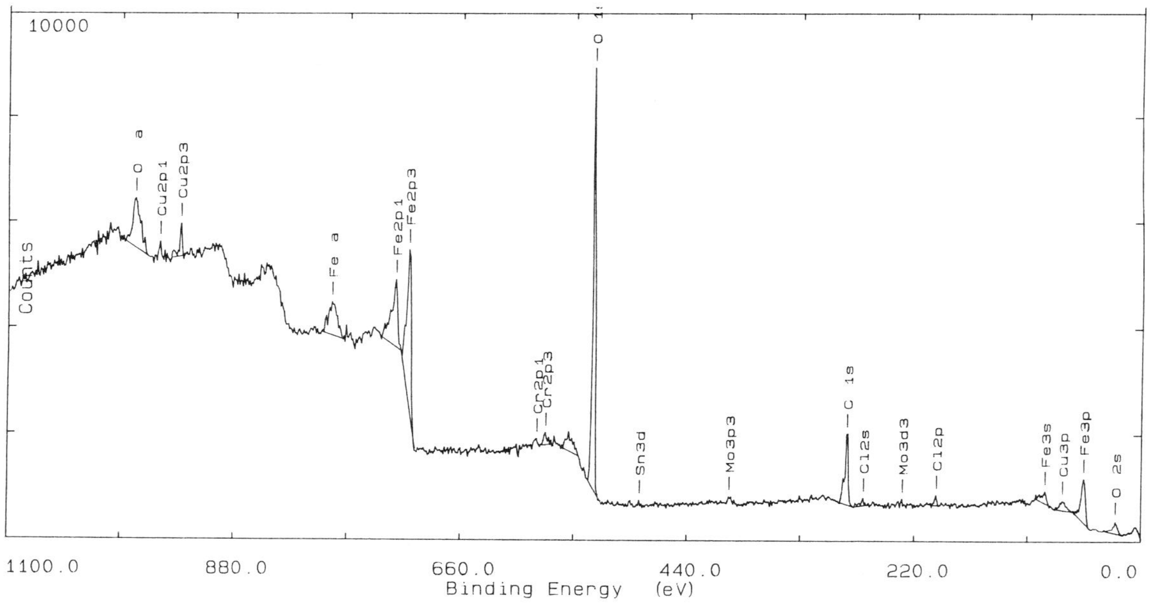

Composition table

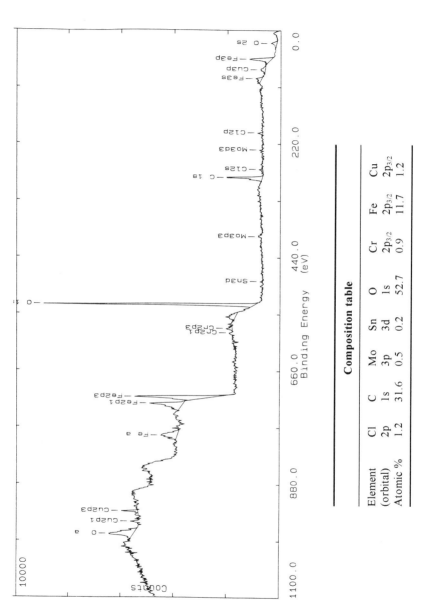

Element (orbital)	Cl 2p	C 1s	Mo 3p	Sn 3d	O 1s	Cr $2p_{3/2}$	Fe $2p_{3/2}$	Cu $2p_{3/2}$
Atomic %	1.2	31.6	0.5	0.2	52.7	0.9	11.7	1.2

Figure 13.2 An XPS survey spectrum and composition table recorded from a coupon of steel taken from a highly corrosive environment.

initial core electron level hole and this can be generated by an electron beam, X-rays, ions or even by thermal energy. The exact nature of the primary excitation source on the sample is not important. The characteristic kinetic energy of the Auger electron is only dependent on the binding energies of the core levels within the target atom itself, as in Equation 13.2 and Fig. 13.1 (b).

$$E_K = E_{B1} - E_{B2} - E^*_{B3} \qquad [13.2]$$

where: E_{B1} = the binding energy (B_E) of the core hole (eV)
E_{B2} = the *BE* of the electron which drops to the core hole
E^*_{B3} = the BE of the electron level from which the Auger electron is ejected. The asterisk indicates that the binding energies have moved to higher values due to the reduced charge screening associated with the loss of the original core electron.

Classic X-ray notation is used to specify the Auger bands; for example a [1s, 2p, 2p] is designated $KL_{2,3}\,L_{2,3}$ whilst [2p,3p,3d] is known as $L_{2,3}\,M_{2,3}\,M_{4,5}$.

Auger electron bands are observed in photoelectron spectra as seen in Fig. 13.2. However, *Auger electron spectroscopy* (AES) has been studied traditionally by excitation with a primary electron beam. As a result, *Auger electron spectroscopy* has been combined with electron microscopy to produce the powerful technique of *scanning auger microscopy*, SAM.

13.2.3 XPS, AES: chemical information

Atomic orbitals from atoms of the same elements in different chemical environments possess slightly different but measurable binding energies. This effect is in the range 0.1–10 eV and is described [9] as the *chemical shift*.

Chemical shifts arise because of the variations in electrostatic screening experienced by core electrons as the valence and conduction electrons are drawn towards or away from the specific atom. Differences in oxidation state, molecular environment or coordination number will all provide different chemical shifts.

In principle, AES is also capable of providing information on the chemical state of the element. In practice the chemical shifts are small compared to the broader widths of Auger peaks and this reduces the quality of the chemical information. Nevertheless, distinct changes in Auger peak shape are observed in some cases for different chemical states.

X-ray excited Auger peaks can be recorded at high energy resolution which allows both accurate determination of the exact peak kinetic energies and the fine structure. Chemical shifts in Auger peaks often can be larger than those of the associated photoelectron peaks. This is because the final state of the Auger process is doubly ionized ion and with the local field dependent on the $(\text{charge})^2$, a four fold change in extra atomic relaxation can $DR(K^+)$ occur.

Two types of chemical shifts have been used successfully in the formation of chemical state plots by Wagner et al [10]. For example, zinc shows a much larger shift in the Auger LMM peak than in the corresponding 2p photoelectron peak. Metals and alloys exhibit high values in the Auger peak as conduction electrons contribute to screening and extra atomic relaxation energy. When zinc is bonded to more electronegative species, the screening effect of valence electrons is reduced and along with extra atomic relaxation, there is a shift in both peaks. The diagonal axes are lines of constant *Auger parameter* which is a direct measure of the extra atomic relaxation energy. The Auger parameter is used widely for its chemical sensitivity. It also enjoys a freedom from sample charging problems, as peak separations, and not absolute energies, are measured.

Photoelectron binding energy shifts are nevertheless the principal source of chemical information and have been used in every area of materials science. Measurement of the chemical shift can be of particular value in surface engineering situations where it is important to establish the presence and nature of a surface oxide of a given material or coating for example to explain poor solderability, adhesion or corrosion resistance.

Recently, the introduction of monochromatized X-ray sources in XPS has provided greater energy resolution enabling much more accurate measurement of chemical shifts and improvements in understanding combined with software developments have allowed the proportions of each state to be quantified.

The example in Fig. 13.3 shows a high resolution iron spectrum recorded from the steel sample shown in Fig. 13.2. The high resolution iron spectrum shows the presence of metallic iron, iron carbides, iron oxides and iron carbonate. The peak deconvolution enables the relative proportions of each state to be determined.

13.2.4 Surface sensitivity in the XPS/AES

Although X-rays penetrate deeply (up to several microns) within a solid, XPS and Auger electrons only escape without energy loss from the outermost atomic layers of the sample. This is because the electrons are of low energy and consequently have a very short *inelastic mean free path* (IMFP).

The IMFP is dependent on electron energy and density and nature of the material through which the electrons pass. Seah and Dench [11] derived a series of equations from plots of IMFP against electron kinetic energy on log log scale [1]. In the energy range of XPS and AES (100–2000 eV) , electron IMFP in all solids is between 1 and 10 monolayers which is equivalent to 3–30 Å. An approximate rule of thumb over this range is that:

$$\text{Å} = \frac{1}{2}\sqrt{E} \qquad [13.3]$$

Virtually all the signal observed (i.e. 95%) in XPS and AES is derived from the top 3*l* within the solid. This is dependent also on the *electron take-off angle*

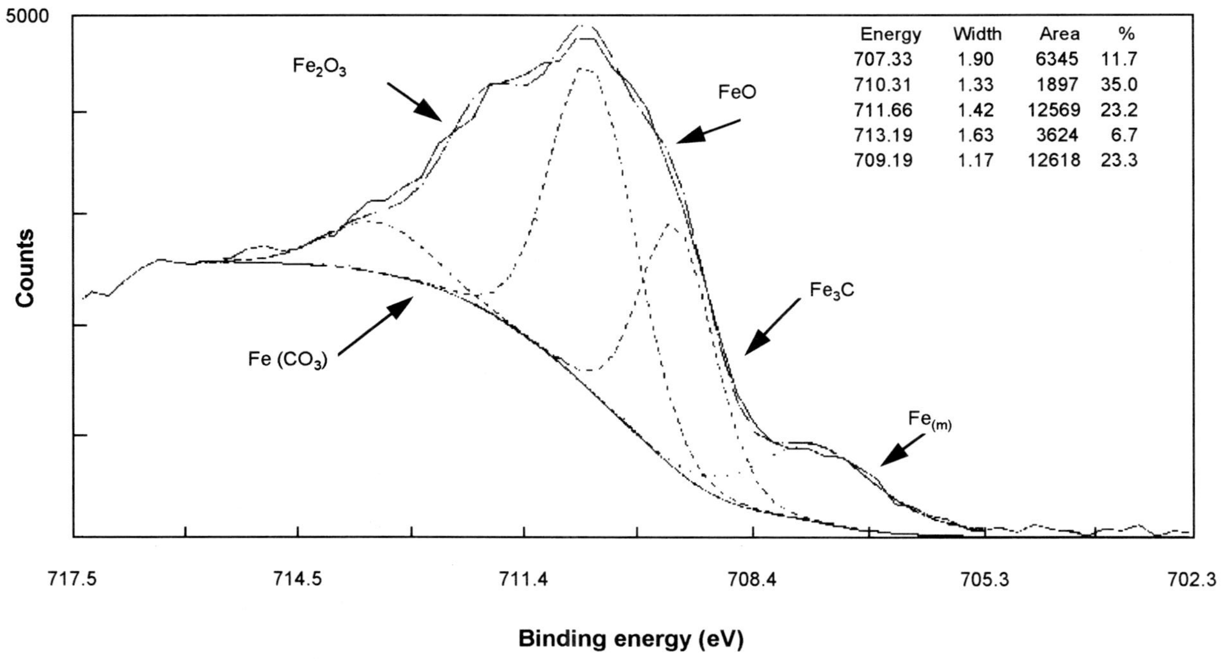

Figure 13.3 The high resolution iron spectrum recorded from the corroded steel specimen in Fig. 13.2.

so that the sampling depth d is defined as:

$$d = 3l \sin \alpha \qquad [13.4]$$

where α is referenced to the sample surface.

The potential for varying surface sensitivity of these techniques is used in the study of thin overlayers on substrates as in Fig. 13.3. Lubricant films, biocompatible coatings and protective oxide films for example often have thicknesses that lie within the régime accessible to angle-dependent XPS. The signal I_s, detected from the substrate is attenuated by the overlayer, as predicted from Beer-Lambert law:

$$I_s = I_s e^{-\frac{d}{l \sin \alpha}} \qquad [13.5]$$

While the signal from the overlayer I_i is dependent mostly on the layer thickness:

$$I_i = I_i (1 - e^{-\frac{d}{l \sin \alpha}}) \qquad [13.6]$$

By monitoring the XPS spectra at varying take-off angles it is possible to study changes in surface composition and chemical state at different depths and so to determine surface structure.

In the example shown in Fig. 13.5, a section of video tape has been examined at two take-off angles. The tape consists of a metal oxide (magnetic memory) layer lying beneath a thin film of fluorocarbon lubricant.

At the shallow take-off angles (15°) strongest signals (fluorine, oxygen and carbon) originate from the fluorocarbon lubricant. (NB; even under the low energy resolution conditions of a survey scan some effect of fluorine atoms of the lubricant on the chemical shift of carbon and oxygen is resolved in the peak splitting seen for these elements). At a steeper take-of angle (90°) a much stronger signal from

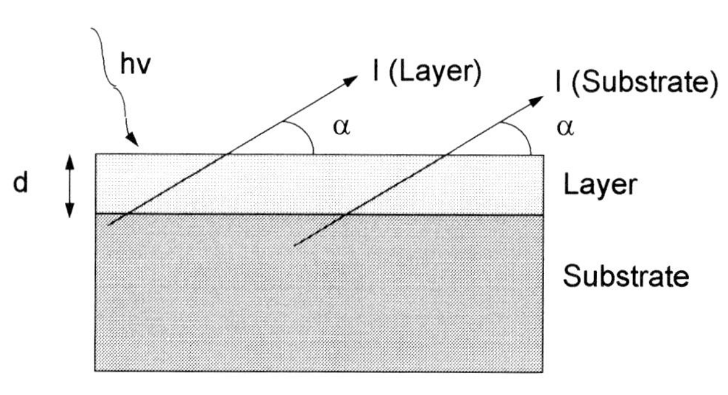

Figure 13.4 Schematic diagram showing variation in information depth with take off-angle (TOA) of photoelectrons.

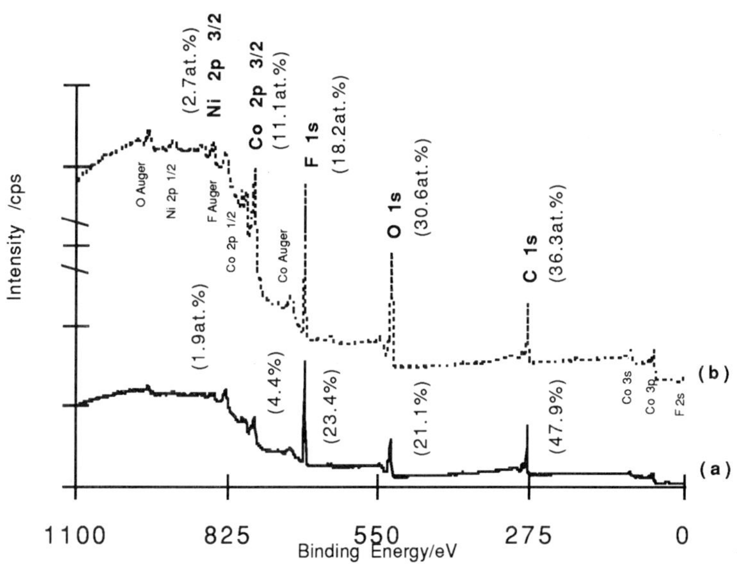

Figure 13.5 XPS survey spectra and composition of video tape recorded at two take-off angles, (a) 15° (minimum sampling depth) and (b) 90° (maximum sampling depth).

the metal oxide-containing layer is detected (cobalt, nickel and oxygen). Furthermore the relative intensities of the fluorinated carbon and oxygen species are seen to diminish with the reduction in surface sensitivity.

For films in the thickness range 1 to 10 nm this experimental approach can be used to calculate film thickness with reasonable accuracy. The technique can also be used to determine the orientation of certain molecules at surfaces.

13.2.5 Spatially resolved information

Both Auger and, more recently, XPS can be used in a '*microprobe*' mode to examine localized features at surfaces. Currently best spatial resolution is achieved by Auger although, imaging, XPS can cope with a wider range of sample types including insulators.

13.2.6 Scanning Auger microprobe (SAM)

Imaging in electron spectroscopy has been dominated by *scanning Auger microscopy* where the primary electron beam is scanned over the surface and the electron analyser is tuned to a specific Auger peak. The detection of Auger electrons illuminates the imaging screen and creates a map of the surface concentration of the element of interest. The applications of this technique are immense in the areas of metallurgy, corrosion and semiconductors, however, electrostatic charging interferes with its application in the study of ceramics and polymers. In addition, beam damage effects are generally greater with electrons than X-rays and become particularly important when small areas are analysed at very high spatial resolution (current density).

Spatial resolution in SAM is limited by the primary beam diameter and the extent of sub-surface interactions. For a 50 nm electron beam, SAM resolution can be as low as 200 nm for favourable samples.

13.2.7 Imaging XPS

The X-ray beam has neither charge nor mass and for this reason it cannot be focused or scanned in the same way as the electron beam for Auger (or the ion beam for SIMS). This phenomenon has, until recently [12], restricted XPS analysis to the analysis of large areas and accelerated the development of micro-imaging capabilities of the aforementioned techniques.

Recent instrumental developments however, render it possible to perform XPS mapping analysis either by scanning the *sample* under a small spot X-ray beam (150 μm), by adapting the selected analysis area technique (100μm) or, most recently, by using a photoelectron microscope arrangement to image the position of surface species by imaging the emitted photoelectrons (10 mm). In any case spatial resolutions do not compare well with those currently attainable using other surface imaging techniques and imaging XPS analysis can be extremely slow. However, there are certain cases where XPS imaging can be valuable. For example, imaging XPS analysis can generally operate with a wider range of more awkward samples, some of which are extremely difficult for scanning SIMS or SAM analysis (such as electrical insulators or very rough surfaces). Different oxidation states of the same element can also be imaged in XPS. It is also highly advantageous that the XPS image is directly quantifiable, since in principle, it means that the full surface composition is available for any image point as well as the mean surface composition over preselected entire imaged areas.

Figure 13.6 shows an XPS image of a failure interface of an aluminium beverage can where a polymeric coating has peeled away from the internal surface of the can. The delamination region can be readily seen in the images for aluminium, oxygen and sulphur (originating from a surface treatment chemical used in can processing).

13.3 Mass spectrometry of surfaces

Mass spectrometry provides detailed information about chemical and structural properties of unknown compounds in analytical chemistry. In the same way *surface* mass spectrometry is of key importance for surface analysis of materials. High resolution mass spectra can be generated from the outermost surface of all types of solid (and some liquid) specimens yielding complementary information to that of electron spectroscopies and rendering this an important technique for understanding the chemical characteristics of surfaces. In addition, surface mass spectrometry is a micro-probe technique, enabling surface and near-surface microanalysis of materials. Such versatility is of key importance to work in *surface engineering* and failure analysis.

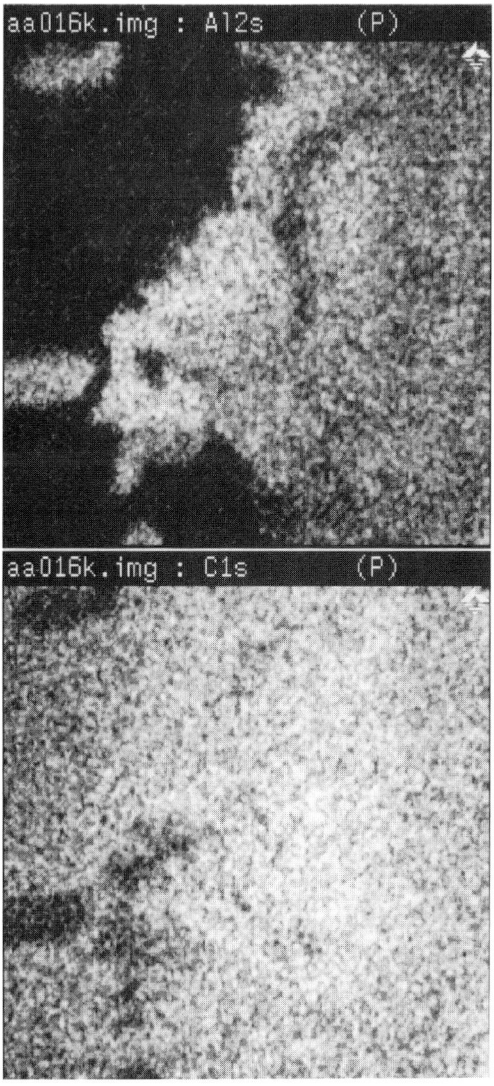

Figure 13.6 An XPS image of the aluminium side of a failed coating/can interface of an aluminium beverage.

Surface mass spectrometry techniques are well suited to the investigation of organic and inorganic materials. Careful selection of the appropriate method and experimental conditions enables rough, powdered, chemically fragile, insulating and other "difficult samples" to be analysed; often with little or no sample pre-treatment. Increasingly, as with ESCA techniques, samples (product) taken directly from in-service situations can be analysed.

The main method of signal generation is by sputtering of ionized particles from the specimen surface. Mass discrimination can involve the use of alternative mass spectrometers which are specially configured for surface analysis. In this section essential features of the sputtering process will be described followed by consideration of instrumental features.

Although the emission of secondary ions from surfaces was observed almost 100 years ago [13], only in the last decade has the analytical potential of surface mass spectrometry begun to be realized; principally through secondary ion mass spectrometry (SIMS). A number of books describe principles and applications of the technique [4,14] and others provide catalogues of reference spectra as handbooks to aid in data interpretation [15].

13.3.1 Signal generation: sputtering

In the sputtering process the specimen surface is bombarded by a primary beam of particles of fixed energy. The particles (commonly ions or atoms), induce a series of hard sphere collision cascades along pathways up to 10–25 nm into the surface region whereupon the majority of energy transfer in this process is absorbed in the surface layers. When the collision cascade returns to the outermost surface and is of sufficient energy to break chemical bonds then atomic and clustered fragments of surface molecules are ejected into the vacuum. There are many explanations of sputtering phenomena, most of which have been comprehensively reviewed and mathematically modelled, e.g. [16].

The sputter yield (Y) measures the number of emitted species (I_S) per number of incident particles (I_p).

$$Y = \frac{I_s}{I_p} \qquad [13.7]$$

The sputter yield varies from element to element and, for the same element, from matrix to matrix generally lying between 0.1 and 10 for most materials; it is also affected by the energy and atomic mass of the primary beam and its angle of incidence to the sample surface. A schematic diagram of the sputtering process is shown in Fig. 13.7.

Two important aspects of sputtering are that; firstly, the energy of collision which results in emission of secondary species is much lower than that of the energy of the primary/impact beam, and secondly, the point of ejection of the secondary particle is remote from the point of impact. For these reasons surface mass spectra can contain highly complex structural information where molecular ions are detected directly.

13.3.2 Secondary ion mass spectrometry (SIMS)

Over 99% of species ejected in the sputtering process are electronically neutral (see section 13.3.7 on SNMS), however, a finite portion become ionized in the emission process. In SIMS analysis the ionized portion of emitted species is extracted directly for analysis in the mass spectrometer. By altering the polarity of extraction, both positive and negative secondary ions can be drawn into the mass spectrometer for analysis.

Secondary ion signal intensity or yield (i_S^M) is a function of the primary ion current I_p, bombarded area A, sputter yield (Y), ionisation probability ($R^{+/-}$), coverage ($\varnothing$) and instrument transmission (n) (predominantly mass spectrometer controlled) as given in Equation [13.8] which shows the secondary ion yield ($i_S^M = I_p$ etc..) for positive ions:

$$i_S^M I_p A Y R^{+/-} \varnothing_M n \qquad [13.8]$$

Unlike ESCA, SIMS is not a directly quantitative technique. It can, however, be used accurately in a quantitative manner in conjunction with a series of suitable calibration samples.

13.3.3 Dynamic SIMS vs static SIMS

The secondary ion yield (sensitivity) increases with primary particle flux – Equation [13.8] – but this also increases the amount of material removed from the surface for a given analysis – Equation [13.7].

The lifetime of a surface layer of atoms (t_m in seconds) is equivalent to the number of surface atoms (N_s atoms per square cm) divided by the product of the sputter yield, and the number of incident primary particles ($6.2 \times 10^{18} I_p$ $cm^{-2}s^{-1}$).

$$t_m = \frac{N_s}{6.2 \times 10^{18} I_p Y} \qquad [13.9]$$

Clearly, removal of a large volume of surface atoms from the analysis area during analysis is undesirable for true surface analysis and so minimal primary particle fluxes are preferred. This mode of SIMS is called *static* SIMS; whereby the surface chemical state can be considered to remain unaltered or static as a consequence of analysis.

Alternatively, at higher primary beam fluxes significant amounts of surface

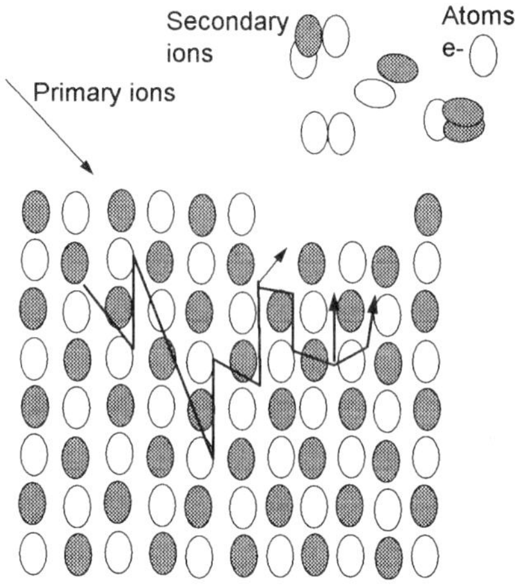

Figure 13.7 Schematic diagram of the sputtering process.

species can be removed at fast rates and the SIMS signal can be monitored from successive layers of a material as a function of removal rate. This mode of analysis is termed *dynamic* SIMS; whereby the surface is eroded in real time as a consequence of analysis. This is the mode of SIMS used for probing near-surface composition by depth profile analysis.

It is possible to perform both static and dynamic SIMS using the same instrument by adjusting the experimental conditions. Practical considerations (such as types of ion source, mass spectrometer, optimization of geometry etc.) have led to the manufacture of dedicated instruments for optimal analysis in either mode.

In the development of the SIMS technique it also became apparent that each of the modes were most useful for different types of material problems presented by industry. Depth profiling SIMS found applications in the semiconductor industry benefiting from a unique capability to chemically identify the ultra low levels of charge carriers in semiconductor materials and to characterize the layer structure of devices. Static SIMS, however, has found a wider range of applications including work on metallic [17], catalyst [18] and semiconductor [19] surfaces in conjunction with other surface analysis techniques, to biomolecular [20] and industrial organic coatings [21]. Sensitivity to molecular structure and capability for characterisation of complex organic surfaces has been found to be of particular benefit for the characterisation of polymer surfaces [22,23]

13.3.4 Static SIMS

Of all the applied surface analysis techniques, static SIMS is the most sensitive to molecular structure. Whilst elemental sensitivity, surface specificity and spatial resolution account for many applications of static SIMS in semiconductor, catalyst and metallurgical fields, it is in the analysis of organic surfaces where static SIMS is of key value and molecular structure interpretation most fully exploited, especially since the introduction of new high powered *time of flight* (ToF) mass spectrometry to SSIMS [24].

Minimisation of surface damage is important to protect the molecular integrity of the surface during analysis and to stabilize/preserve high mass information. A critical ion dose of 10^{-13} particles cms^{-2} has been accepted as the threshold value for primary ion beam-induced damage effects to become evident and for the static SIMS experiment to transgress into the dynamic mode [25]. This threshold value is approached only in exceptional circumstances in SSIMS analysis using modern ToF instruments. In practical terms the static conditions are most limiting when analysis of extremely small areas is required.

Different types of mass spectrometer and ion sources can be used for static SIMS analysis and these have been reviewed elsewhere [4,8]. The latest ToFSIMS instruments offer a mass resolution to the third or fourth decimal place of mass enabling accurate mass measurement and mass permutation calculation to unequivocally identify the elemental combination (empirical formula) of a given peak.

The example shown is of the ToFSIMS analysis of a typical oilfield pipeline corrosion inhibitor; *oleic imidazoline*. Figure 13.8 (a) shows the molecular ion region of the SIMS spectrum where key molecular fragments are identified. As in conventional mass spectrometry, assignment of key cluster or fragment ions is made based upon identification of the molecular ion and low mass key diagnostic ions with reference to existing data [15]. Using the latest high mass resolution ToFSIMS instruments, as in this case, assignment of the molecular ion can be confirmed by accurate mass measurement of the ion followed by a permutation calculation of accurate isotopic masses. (Fig. 13.8 (b)).

This sample was prepared by smearing a thin film of inhibitor (a viscous liquid) on to a clean substrate (aluminium foil), the resolving power and enhanced sensitivity of the latest generation ToFSIMS instruments is revealed at low mass where organic and aluminium peaks at the same nominal mass (27 amu) can be separated. High mass resolution improves the sensitivity of static SIMS towards its theoretical detection limit of ppb for surface elements and widens the applicability of ToFSIMS into the area of *"trace analysis"* which is likely to be of importance for the analysis of ultra-low level dopants in engineered surfaces.

The example shown in Fig. 13.8 also demonstrates how ToFSIMS can be used for the analysis of certain liquids. However, the usefulness of this ToFSIMS is extended when samples are taken from "real world" situations and this approach is generally possible for solids and liquid films after careful thought about sample transfer. Examples can be found in the literature describing the application of ToFSIMS amongst other surface analysis techniques for the characterisation of various lubricant/friction films on components taken from tribological testing situations, biocompatible coatings taken from aqueous/saline or serum solutions, corrosion inhibitor films on metal taken directly from aqueous corrosive environments, and industrial surfacants e.g. [24,26,27].

13.3.5 Dynamic SIMS

When the primary ion dose is increased beyond 10^{-13} ions cm^{-2} s^{-1} the rate of consumption of surface layers in the emission process becomes significant and SIMS enters the dynamic régime. Very often much higher ion doses are used in dynamic SIMS in order to depth profile the uppermost several nanometres–micrometres of the near-surface.

The principal features of dynamic SIMS of practical importance are its sensitivity, depth resolution and dynamic range. In dynamic SIMS analysis the primary beam of ions bombards the surface at a precisely controlled flux and energy. Typically depth profiling SIMS experiments will be performed with a moderately high particle flux and so secondary ion yields are high – Equation [13.8] – and sensitivities reach ppb in favourable cases. Whilst these may be desirable conditions from the point of view of sensitivity, depth resolution and cluster ion stability are forfeited by working at high particle fluxes. This means that often a compromize between sensitivity and depth resolution is selected. In contrast to static SIMS, a depth profile SIMS analysis is mostly concerned with elemental or simple cluster ion distributions with depth.

In surface engineering applications dynamic SIMS is mostly used as a semi-quantitative technique, where, for example in coatings evaluation the purity of a novel coating is assessed against a known reference material. In the example shown in Fig. 13.9 depth profiling SIMS has been used to determine the distribution of chromium in a chromate conversion layer applied to aluminium. The SIMS depth profile shows how chromium is strictly located in the outermost 20Å of coating under which lies the aluminium oxide metal.

For most semiconductor applications it is a requirement that the SIMS analysis is quantitative. In this case since the sample matrices are often very similar for these materials, accurate standards can be prepared and the technique is commonly used to measure dopant concentrations. This can be achieved with an accuracy dependent upon the system under investigation but typically to within ~5%. In many *surface engineering* applications such a high degree of precision in quantitative assessment is not required and it is sufficient to perform comparative analysis (to the ppm–ppb lower detection limit) between unknown and reference samples.

For a given material, the depth scale can also be calibrated accurately by post analysis measurement of crater depth (for example mechanically by surface profilometry or light interferometry). However, this means of depth calibration becomes more difficult and less accurate for complex layer structures of materials with widely different sputter yields and the errors are substantial for very shallow profiles.

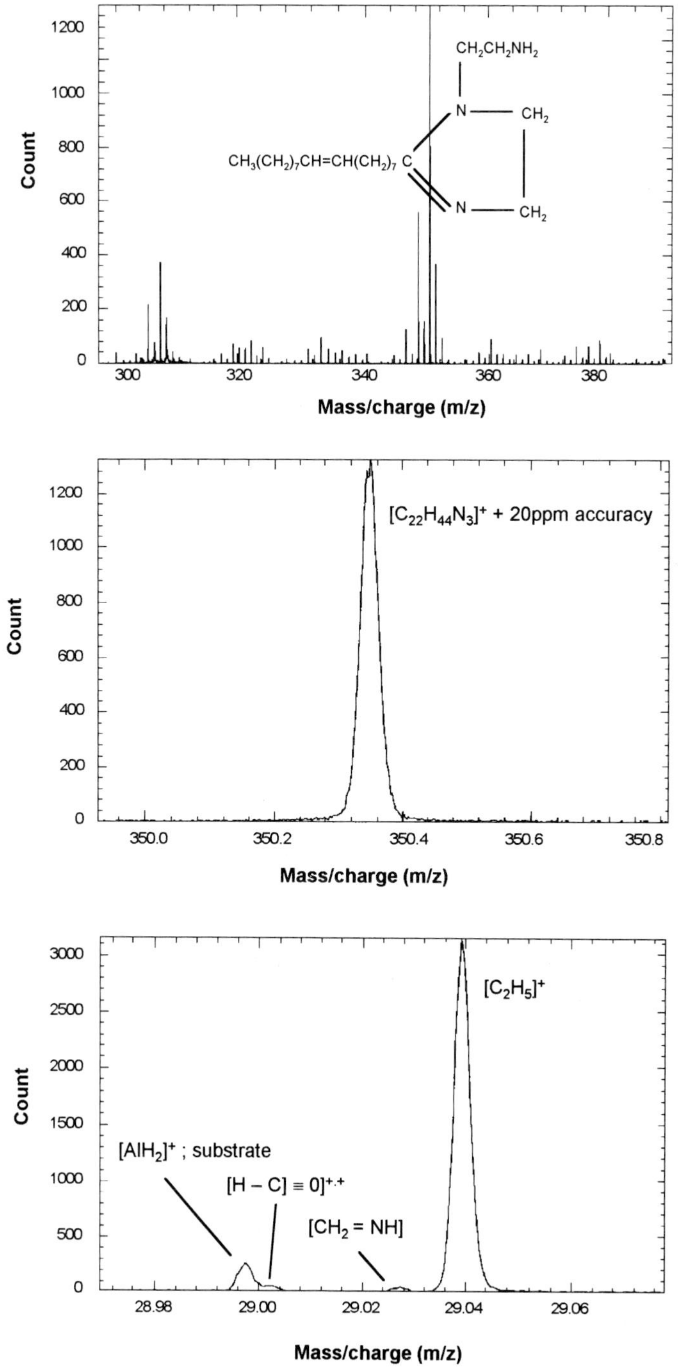

Figure 13.8 Selected regions of the high resolution ToFSIMS spectrum oleic imidazoline corrosion inhibitor.

In depth profiling SIMS, dynamic range is often optimized by electronically or optically gating the detected signal. This is a method by which the signal is selected only from the central area of the etched crater thus discounting the contribution of signal/effects from the crater edges. With electronic gating this is achieved by deadening the counting electronics when the scanning primary beam is at the crater centre by preset coordinates, with optical gating the field of view of analysis is set by control of the ion optics arrangement at the entrance to the mass spectrometer. Some instruments use a combination of both methods of gating of the SIMS signal. It follows that gating can also improve the depth resolution.

Depth profiling SIMS practitioners have developed a number of other practical means to improve these parameters not all of which are by instrument control [28].

13.3.6 Chemical mapping – imaging SIMS

In static and dynamic SIMS the primary ion beam cam be raster-scanned across the surface region of interest. Scanning the primary ion beam with the mass spectrometer set to detect preselected secondary ions enables the distribution of these species to be mapped over the area of interest. The spatial resolution attainable in this mode is primarily controlled by the beam diameter and the sensitivity per pixel of the instrument for the species of interest. Best spatial resolutions are achieved using *microfocused (primary) ion guns* (MIGs) which offer narrow spot size and high *"brightness"* (high ion current per unit area). A variety of MIG designs exist which generate primary ions from a reservoir of liquid metal (most commonly indium, gallium and caesium). Analysis using the so-called (*scanning*) SIMS microprobe uses computer image storage and colour coded graphics systems to produce colour coded maps of the relative intensity of elemental and molecular ions in the surface.

An alternative method of mapping the distribution of surface elements is derived from using an unscanned primary ion beam and a mass spectrometer with an ion optic arrangement set up so that the positional sense of the ions is retained throughout mass discrimination. This mode of mapping is known as *ion microscopy* where a virtual image of secondary ion emission is displayed directly on to a fluorescent screen or registered on a position sensitive detector for subsequent computer storage/manipulation.

There are many applications of imaging/scanning SIMS in surface engineering situations where the technique can be used to visually monitor the integrity of a functional coating or surface treatment. The example shown in Fig. 13.10 is of a cross-section of a glass fibre reinforced composite material analysed by scanning SIMS [29]. The composite has been sectioned and polished before analysis using a MIG on an instrument fitted with a quadrupile mass spectrometer. Fibres are clearly distinguished by the map of oxygen secondary ion signal red/orange originating from the glass whilst the resin distribution is conveniently mapped on

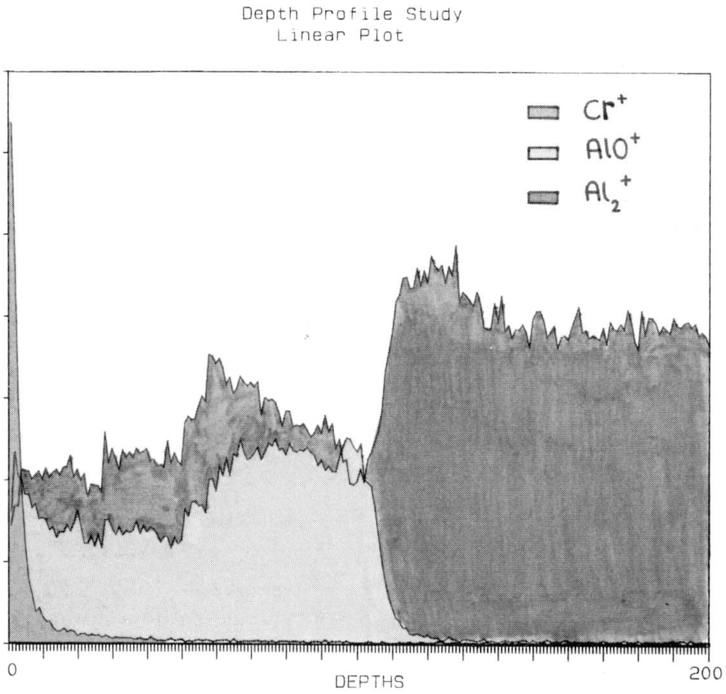

Figure 13.9 A SIMS depth profile of the distribution of chromium, aluminium oxide and aluminium species throughout the outermost 360Å of a chromate conversion layer applied to aluminium.

the CN-signal. In this work the distribution of a chlorine-containing coupling agent was of key interest. The corresponding chloride ion map shows that, in this case the coupling agent is located at the fibre/resin interface and partially into the resin itself. This is confirmed in the retrospective linescan of signal intensity (same colour coding) shown.

Critical features of scanning/imaging SIMS analysis are often: (i) the capability to acquire an image under static conditions, and (ii) the ability to map the distribution of molecular (or other high mass) ions. Both aspects relate to the potential of the technique to map true surface chemical features without imparting *"significant damage"* to the surface and both are best achieved using the *time of flight mass spectrometry*.

13.3.7 Sputtered neutral mass spectrometry (SNMS)

In SNMS an attempt is made to ionize the portion of neutral species which are emitted in the sputtering process. Since these species constitute the vast majority of the sputtered *"plume"* of particles ejected from the surface, then in principle,

SNMS should be able to provide surface mass spectral information which is more quantitative than SIMS. Whilst quantification is improved in SNMS, accuracy remains a problem since rejection of secondary ions (whose intensity is not solely a function of surface concentration – Equation [13.8] – is not complete.

However, since sputter-atom yields vary only by ~3–5% for the same element in different matrices, then matrix effects are much less evident in the SNMS process. The technique also affords higher sensitivity than SIMS for analysis of those species of low ionisation potential. Both features are particularly advantageous in depth profiling SIMS analysis where the requirement for high sensitivity to all elements is premium and changes in elemental profiles can be severely distorted by changes in matrix from layer to layer. Indeed, it is in depth profiling analysis where SNMS has its majority of applications since, generally, a sufficient yield of neutrals is only achieved under *"dynamic"* conditions.

13.3.8 Laser desorption mass spectrometry (LDMS, LIMA)

Laser light can also be used as the primary source for the generation of mass spectra from the surfaces of materials in a technique known as *laser ionisation mass analysis/spectrometry* (LIMA/S). This can be an extremely rapid and convenient method for acquiring near-surface mass spectral information from small areas at high sensitivity.

In this technique a high energy pulsed UV or visible wavelength laser is used to absorb and ionize the emitted species in a one step process followed by ToF mass analysis of the ensuing spectrum. Since many effects occur at the material surface under high energy laser exposure, the ion emission process is much less well understood compared to sputtering SIMS. A combination of thermal and resonance effects is thought to be the principal activator for emission. Coupling of the laser wavelength with the matrix of the material under investigation is required.

A LIMA spectrum generally contains much fewer high mass molecular ions than the corresponding SSIMS spectrum, however, some lower mass fragment ions are often (but not always) common to both techniques. LIMA analysis is much less *surface* specific than static SIMS and is variable with the thermal properties of the sample. Surface specificity can be improved by performing the analysis at shallow angles of incidence of the laser light.

13.4 Summary

As the technology of modern materials advances, the development and characterisation of precisely engineered surfaces is set to be of increasing importance. This is the case for an ever-widening range of industries from, for example, packaging to the manufacture of surgical appliances.

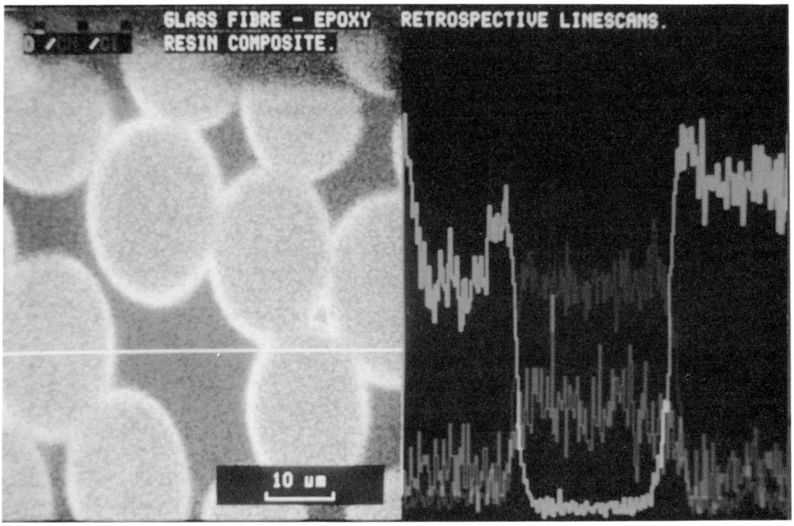

Figure 13.10 Scanning SIMS map of a cross-section of a glass fibre reinforced composite material.

Methods of surface engineering for all types of material have developed to such levels of precision and complexity that traditional means of "near-surface" analysis such as electron microscopy are often no longer adequate to characterize the engineered surface or to diagnose failure. Instead advanced techniques which can probe the outermost atomic layers of materials with high sensitivity to all of the elements in the periodic table, sensitivity to molecular structure of organic materials and which can depth profile or map distribution of key species in or across the near-surface regions are required.

This Chapter has shown how a combination of electron spectroscopy and surface mass spectrometry can meet many of these challenges for understanding the composition and chemical/molecular state of surface and near-surface regions of all types of materials.

References

1. Briggs D and Seah M P, *Practical Surface Analysis by Auger and X-ray Photoelectron Spectroscopy*, 2nd Edition, Volume 1, J. Wiley, Chichester, 1990.
2. Holt D B, Muir M D, Grant P R and Boswarva I M, *Quantitive Scanning*

Electron Microscopy, Academic Press, London, 1974.

3. *Practical Scanning Electron Microscopy*, Plenum Press, New York & London, 1975.
4. Eds. Vickerman J C, Brown A and Reed N M, *Secondary Ion Mass Spectrometry Principles and Applications*, Int. series of Monographs on Chemistry, 17, Oxford University Press, 1989.
5. Smith G C, *Quantitative Surface Analysis for Materials Science*, Inst. Materials, 1991.
6. Walls J M (Ed.) *Methods of Surface Analysis*, Cambridge University Press, Cambridge, 1989.
7. Swift A J and West R H, Chapter 10, Advanced Instrumental Analysis, Ed. R. Denney, Spektrum Academic Publishing, 1994.
8. Eds. Vickerman J C and Reed N M, *Surface Analysis and its Applications*, John Wiley and Sons,1994.
9. Siegbahn K et al, *Atomic, Molecular and Solid State Structure Studied by means of Electron Spectroscopy*, Almquist & Wiksells, Uppsala, Sweden, 1967.
10. Wagner C D, Gale L H and Raymond R H, *Analytical Chemistry*, **51**, P446, 1979.
11. Seah M P and Dench W A, *Surface Interface Anal.* **1**, P2, 1979.
12. Drummond I W, *Microscopy and Analysis*, march, P29, 1992.
13. Thompson J J, *Philos. Mag.*, **20**, 252, 1910.
14. Benninghoven A, Rudenauer F G and Werner H W, *Secondary Ion Mass Spectrometry, Basic Concepts, Instrumental Aspects, Applications and Trends*, Monographs Ion Analytical Chemistry and its Applications, John Wiley and Sons, p86, 1987.
15. Briggs D, Brown A, and Vickerman J C, *Handbook of Static Secondary Ion Mass Spectrometry*, John Wiley and Sons Ltd., Chichister /New York, 1989.

or Newman G, Carlson B A, Michael R S and Hohlt T A, *Static SIMS Handbook of Polymer Analysis*, Perkin-Elmer Corporation, Physical Electronics Division, 6509 Flying Cloud Drive, Eden Prairie, Minnesota, 1991.

16. Sigmund P I, *Sputtering by Particle Bombardment*, in Topics in Applied Physics, R. Behrish (Ed.), Springer-Verlag, Berlin, **47**, p9, 1981.

or Armour, et al., *Nucl. Inst. and Methods in Physics*, **B64**, p609, 1992.

17. Lees D G and Johnson D, *Oxidation of Metals*, in press, 1992.
18. Sakakini B, Swift A J, Vickerman J C, Harendt C and Christman K, *J. Chem. Soc., Farad. Trans I*, **83**, 1975, 1987.
19. Brown A, Humphrey P and Vickerman J C in *Proc. SIMS VI*, 393 in [6].
20. Davies M C, Brown A J, Newton M and Chapman S R, *Surf. and Intl. Anal.*, **11**, p591.
21. van Ooij W J, Sabata A and Appelhaus A D, *Surf. and Int. Anal.*, **17**, p403, 1991.
22. Briggs D, *Polymer*, **25**, p1379, 1984.

or Hearn M J and Briggs D, *Surf. and Interf. Anal.*, **11**, p198, 1988.
23. Gardella J A, Novak F and Hercules D M, *Anal. Chem.*, **56**, p1371, 1984.
24. Swift A J, *Microkimica Acta*, submitted April 1994.
25. Briggs D and Wooton A B, *Surf and Interf. Anal.*, **4**, 3, p109, 1982.
26. Paul A J and Vickerman J C, *Phil. Trans. R. Soc. Lond.* A **333,** p147, 1990.
27. Delargy K, Seeny A and Bell J, Wear. Proc. Conf. Particles, Leeds, 1980.
28. Von Creigern R, Depth Profiling Difficult Samples, Meeting Review UK SIMS Users Forum Newsletter Ed. A J Swift, **3**, 1989.
29. By kind permission of Shell research, KSLA.

Index

C

D

E

F

J

K

L

M

N

O

P